GUIDE TO DUST EXPLOSION PREVENTION AND PROTECTION

PART 3 – Venting of Weak Explosions and the Effect of Vent Ducts: A British Materials Handling Board Design Guide for Practical Systems

Copyright © 1988
THE INSTITUTION OF CHEMICAL ENGINEERS

All rights reserved. No part of this publication may be reproduced, stored in a retrieval system, or transmitted, in any form or by any means, electronic, mechanical, photocopying, recording ot otherwise, without the prior permission of the coyright owner.

ISBN 0 85295 230 9

This Guide is published and recommended by the Institution of Chemical Engineers as a valuable contribution to safety. The information in the Guide is given in good faith and belief in its accuracy, but does not imply the acceptance of any legal liability or responsibility whatsoever, by the Institution or by the author for the consequences of its misuse in any particular circumstances.

1988 Edition

PREFACE

This is the third part of the guide. It consists of the British Materials Handling Board Design Guide on both the effect of vent ducts on the reduced explosion pressures of vented dust explosions, and estimating the venting requirements for weak dust explosions.

Part 1 of the guide dealt with venting and Part 2 with ignition prevention, containment, inerting, suppression and isolation. The guide aims to help those responsible for the design, supply and operation of plant to comply with the provisions of the Health and Safety at Work Act and the Factories Act.

It must be recognised that on occasions strict adherence to these recommendations would be inappropriate and further advice may have to be sought. In addition, it would be expected that further research and other developments will lead to improved methods and it is not the intention of this guide to inhibit such developments.

The material for this book resulted from an extensive dust explosion ducting project funded jointly by the Government and individual companies and run by the British Materials Handling Board.

When $P_{red} > 2.0$ bar abs and choked flow occurs in the vent, an alternative equation is:–

$$A = \frac{V_L^{1/3} \, V^{2/3} (dP_{ex}/dt) \, P_{red}, \, V_L}{\gamma \left(\dfrac{2}{\gamma+1}\right)^{\frac{1}{\gamma-1}} \left(\dfrac{2RT}{M} \dfrac{\gamma}{\gamma+1}\right) Cd \, P_{red}}$$

where γ = specific heat ratio.

Equation [4] is the equation used to derive the K_{st} Nomographs published in VDI 3673. Experiments by Donat yielded values of $(dP_{ex}/dt) \, P_{red}, \, V_L$ for vent bursting pressures of 1.1, 1.2 and 1.5 bar abs. If V_L is taken as 1 m^3, then $(dP_{ex}/dt) \, P_{red}, \, V_L$ can be written as equal to $B \times K_{st}$, where B is a factor with the following values: for $P_{stat} = 1.1$ bar a, $B = 0.283$; for $P_{stat} = 1.2$ bar a, $B = 0.3333$ and for $P_{stat} = 1.5$ bar a, $B = 0.50$. Substituting these figures into Equation [4] gives vent areas in agreement with predictions from the K_{st} Nomographs. Furthermore, the relationship between $(dP/dt)_{max}$ or K_{st} value and $(dP/dt)_{P_{red}}$ can be taken to be linear in the range $K_{st} = 0$ bar m s^{-1} to $K_{st} = 300$ bar m s^{-1}, and thus the values of B can be taken to be a function of P_{stat} only. Thus an extrapolation to K_{st} values less than 50 bar m s^{-1} becomes straightforward. In Figures 1, 2 and 3 the original K_{st} Nomographs are extended, using Heinrich's equation, down to K_{st} values of 10 bar m s^{-1}.

How the Nomographs are used is shown diagramatically in Figure 1.

A series of experiments has been carried out to validate the extended Nomographs. They show that it is important to measure the maximum value of K_{st} when using low K_{st} dusts. Several measurements of K_{st} should be taken, at the most explosible dust concentration, and the highest value measured used when determining either P_{red} or the vent area by the extended Nomographs.

Used directly the extended Nomographs give a low safety factor with low K_{st} dusts. It is recommended that an extra safety factor be incorporated by multiplying the estimated vent area by 1.5.

A GUIDE TO DUST EXPLOSION, PART 3

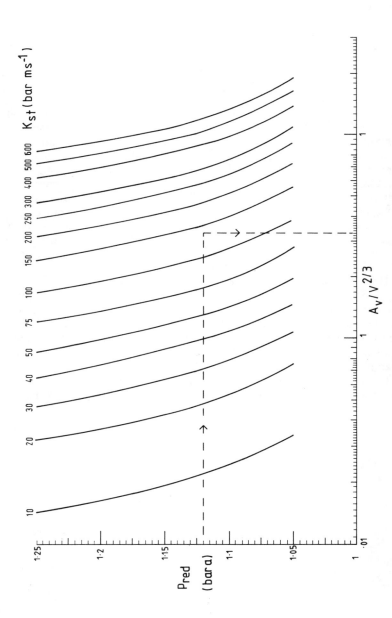

Figure 4. Extended Nomograph: $P_{red} < 1.2$ bar a against $A_v/V^{2/3}$. Estimation of vent areas for $P_{red} < 1.2$ bar a.

2. VENTING REQUIREMENTS WHEN THE VESSEL STRENGTH IS LESS THAN 1.2 BAR A.

Figure 4 gives design information for estimating reduced explosion pressures as a function of the ratio (vent area/(vessel volume)$^{2/3}$), or vice versa, to obtain a reduced pressure of less than 1.2 bar a.

Figure 4 makes no allowances for the value of P_{stat}. It is recommended that in the range of reduced explosion pressures 1.05 bar a $< P_{red} <$ 1.2 bar a, the value of P_{stat} should not exceed $1. + \dfrac{(P_{red}-1.)}{2}$ bar a.

When vent areas are reduced ($<.1/m^2$) then the pressure at which the vent becomes effective becomes of increasing importance. When small vent areas are used the opening pressure should be accurately known and maintenance be such as to maintain this value.

A worked example:

A dust collector with a dirty side free volume of 9.8 m^3 is able to withstand safely a reduced explosion pressure of 1.12 bar a. It will be handling a dust with a K_{st}-value of 135 bar m s^{-1}. Calculate the vent area necessary to limit the explosion pressure to 1.12 bar a, and specify an upper limit to the vent opening pressure, P_{stat}.

ANSWER:
From Figure 4:

At P_{red} = 1.12 bar a, and K_{st} = 135 bar m s^{-1}:

$$A_v/V^{2/3} = 0.33$$

Therefore

$$A_v = 0.33 \times 4.58 = 1.51 \text{ m}^2$$

The vent bursting pressure, P_{stat}, should not exceed $1 + \dfrac{(1.12-1)}{2} = 1.06$ bar a.

Vent Area = 1.51 m^2
Vent Bursting Pressure should not exceed 1.06 bar a

This example is demonstrated in Figure 4.

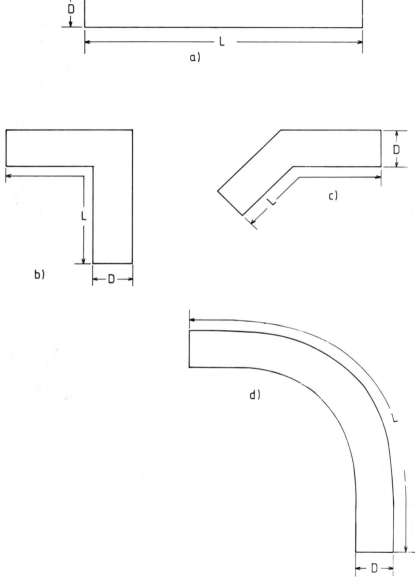

Figure 5.

Measuring the Vent duct L/D ratio.
a) Straight duct (see Appendix 1) b) Duct with a single sharp 90° bend (see Appendix 3) c) Duct with a single sharp 45° bend (see Appendix 2) d) Duct with a gradual bend ($\frac{\text{axial radius of bend curvature}}{\text{duct diameter}} > 2$) (see Appendix 1)

PART TWO: DETERMINING THE EFFECT OF VENT DUCTS ON THE REDUCED EXPLOSION PRESSURE

1. **THE STRUCTURE OF THE DESIGN INFORMATION**

The design guide comprises four sets of graphs:

In APPENDIX 1, information is given on the effect on the reduced explosion pressure of straight vent ducts.

In APPENDIX 2. information is given on the effect on the reduced explosion pressure of vent ducts containing a single sharp 45° bend.

In APPENDIX 3, information is given on the effect on the reduced explosion pressure of vent ducts containing a single sharp 90° bend.

In APPENDIX 4, information is given on the effect on the reduced explosion pressure of straight vent ducts for metal dusts such as aluminium in the St3 group. (300 bar m s^{-1} < K_{st} < 600 bar m s^{-1}).

The preliminary information required for using the guide is the K_{st}-value of the dust (bar m s^{-1}) and the bursting pressure of the vent cover, P_{stat} (bar a). Each Appendix contains a graph applicable to the K_{st} and P_{stat} values used in the K_{st} Nomographs given in PART ONE [1] of this guide.

These are:

K_{st} values (bar m s^{-1}): 10, 20, 30, 40, 50, 75, 100, 150, 200, 250, 300, 400, 500, 600.

P_{stat} (bar a): 1.1, 1.2, 1.5.

Information for reduced pressures, P_{red}, less than 1.2 bar a is included on the charts for P_{stat} = 1.1 bar a, although when P_{red} is less than 1.2 bar a, P_{stat} should not exceed the value suggested in Part One of the design guide.

The graphs are applicable to vent ducts less than 16 m in length of circular cross-section, with the cross-sectional area equal to the area of the vent, and to vessel volumes less than 1000 m^3. Each graph shows the effect on the reduced explosion pressure of vent ducts characterised by the L/D ratio, beginning at L/D = 0. L is the length of the vent duct and D is the diameter. Methods for estimating L are shown diagramatically in Figure 5. The reduced pressure at L/D = 0, $(P_{red})_o$, has been calculated using Heinrich's equation for vented dust explosions on which the K_{st}-Nomographs are based. The additional information required for this estimation is the vessel volume, V, and the area of the vent, A_V. For given values of K_{st} and P_{stat}, Heinrich's method predicts a $(P_{red})_o$ that is dependent on $(A_V/V^{2/3})$.

$(P_{red})_o$ contains a safety factor already incorporated in Heinrich's method. The calculation used to calculate the effect of the vent ducts does not incorporate any additional safety factor.

When using the design guide it may be necessary to interpolate between graphs and between the lines on any particular graph.

A GUIDE TO DUST EXPLOSION, PART 3

This design guide is for circular vent ducts which do not contain any dust prior to a vented explosion passing through them. It is not applicable in any way to dust-carrying pipelines.

A series of worked examples now follows. Figure 6 demonstrates how to use the graphs in Appendices 1 to 4.

When curves on a particular graph merge, always take the highest value of reduced explosion pressure.

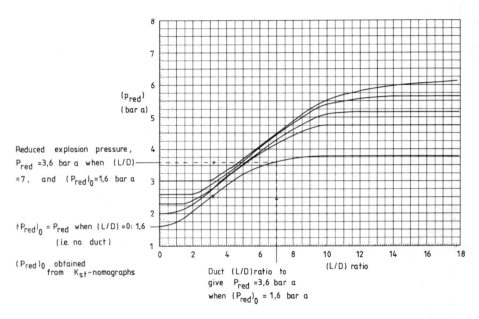

Figure 6. Example of method of using Appendices 1–4

2. HOW TO USE THE DESIGN GUIDE: WORKED EXAMPLES

WORKED EXAMPLE [1]

A vessel of 6 m^3 capacity is used to handle a dust with a K_{st}-value of 150 bar m s^{-1}. A vent with an area of 0.7 m^2 is fitted, closed with a vent panel having an opening pressure of 1.2 bar a. A vent duct 5 m long containing a 45° bend is to be attached to the vent opening. Estimate the reduced explosion pressure in the vessel.

ANSWER

From the K_{st}-Nomograph in Figure 2, find $(P_{red})_o$, the reduced explosion pressure when no vent duct is fitted. $(P_{red})_o = 1.38$ bar a.
Assume 5 m is the length of duct, L, as measured in the method given in Figure 5.
The diameter of the duct, $D = (4 \times 0.7/3.142)^{1/2} = 0.944$ m.
(L/D) ratio = 5/0.944 = 5.3
From Appendix 2, using the graph (Figure A58) for $K_{st} = 150$ bar m s^{-1} and $P_{stat} = 1.2$ bar a:–

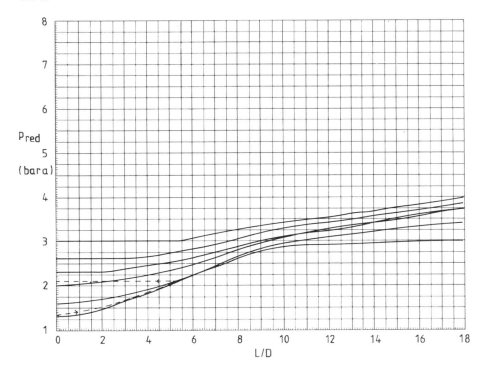

When $(P_{red})_o = 1.38$ bar a and $(L/D) = 5.3$, the reduced explosion pressure = 2.1 bar a.

The Reduced Explosion Pressure = 2.1 bar a

THIS EXAMPLE DEMONSTRATES THE STRAIGHTFORWARD ESTIMATION OF REDUCED EXPLOSION PRESSURE WHEN A VENT DUCT IS ATTACHED.

A GUIDE TO DUST EXPLOSION, PART 3

WORKED EXAMPLE [2]

An item of equipment having a free volume of 16 m³ is being used with a dust of K_{st}-value of 250 bar m s⁻¹. The bursting pressure of the vent closure is 1.5 bar a. The vent area is 1.3 m², sufficient to limit the reduced explosion pressure to 2.5 bar a, according to the K_{st}-Nomographs. If a straight vent duct 6.14 m in length is attached to the vent opening, calculate the new reduced explosion pressure. Determine a vent area which, when fitted with a vent duct 6.14 m long will return the explosion pressure to 2.5 bar a.

ANSWER

The vent diameter, D, $= (4 \times 1.3/3.142^{1/2}) = 1.29$ m.
The vent duct (L/D) ratio $= 6.14/1.29 = 4.76$.

From Appendix 1, using the graph (Figure A34) for $K_{st} = 250$ bar m s⁻¹ and $P_{stat} = 1.5$ bar a:–

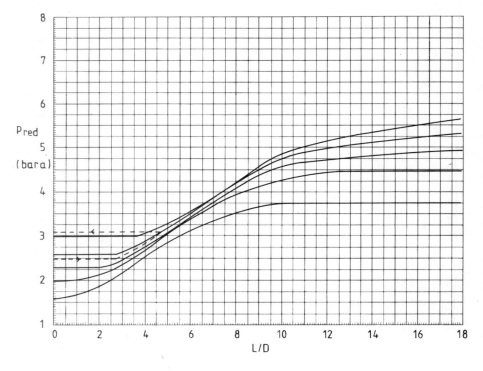

When $(P_{red})_o = 2.5$ bar a and $(L/D) = 4.76$, the reduced explosion pressure = *3.1 bar a.*

To answer the second part of the problem, a trial and error method is necessary, in which a required (L/D) ratio is compared to an actual (L/D) ratio at various values of $(P_{red})_o$. The required vent area is obtained when the two values of (L/D) ratio are equal.

Thus from the same graph in Appendix 1 (K_{st}) = 250 bar m s⁻¹, P_{stat} = 1.5 bar a):–

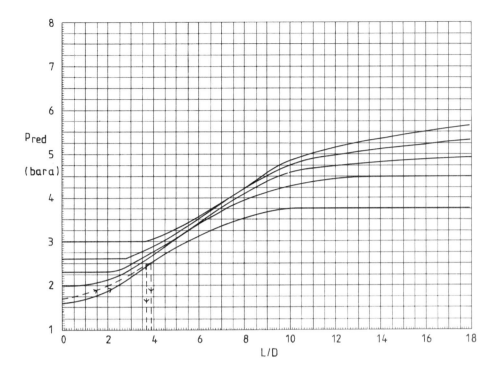

When $(P_{red})_o = 1.6$ bar a, a vent duct with $(L/D) = 3.9$ is required to give an explosion pressure of 2.5 bar a.

From the K_{st}-Nomograph in Figure 3, a vent of area 2.4 m² in a vessel of 16 m³ gives $(P_{red})_o = 1.6$ bar a when $K_{st} = 250$ Bar m s⁻¹. This vent has a diameter $D = (2.4 \times 4/3.142)^{1/2} = 1.75$ m.

Thus the actual (L/D) ratio of the vent duct $= 6.14/1.75 = 3.51$. Which is less than the required (L/D).

When $(P_{red})_o = 1.7$ bar a, reference to the graph from Appendix 1 shows that a vent duct with $(L/D) = 3.7$ is required to give an explosion pressure of 2.5 bar a.

From the K_{st}-Nomograph in Figure 3, a vent of area 2.1 m² in 16 m³ gives $(P_{red})_o = 1.7$ bar a, for $K_{st} = 250$ bar m s⁻¹. This vent has a diameter $D = (2.1 \times 4/3.142)^{1/2} = 1.63$ m.

A GUIDE TO DUST EXPLOSION, PART 3

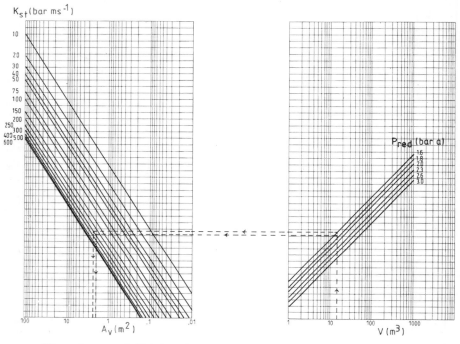

Thus the actual (L/D) ratio of the vent duct $= 6.14/1.63 = 3.77$, which is greater than the required (L/D) ratio.

Using a simple graphical construction:

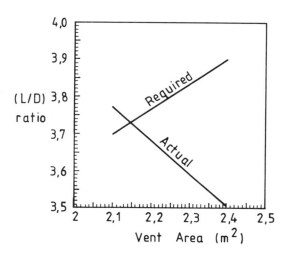

The Required Vent Area = 2.15 m²

This example has been completed by calculating the (L/D) ratios at two values of $(P_{red})_o$ only. It may be necessary, however, to carry out the calculations at more than two values of $(P_{red})_o$, depending on the choice of the initial $(P_{red})_o$, before the final answer can be obtained.

If the dust has a K_{st}–value between two of the graphs in the Appendices, the procedure should be followed at the K_{st}–values immediately above and below the actual value. An interpolation between the vent areas calculated for these two K_{st}–values will then give the required vent area at the known K_{st}–value.

THIS EXAMPLE DEMONSTRATES THE METHOD FOR FINDING A NEW VENT AREA WHEN A VENT DUCT IS ADDED. IN THIS EXAMPLE THE ANSWER LIES WITHIN THE RANGE OF THE GRAPHS. IN WORKED EXAMPLES [3] AND [4] SIMILAR PROBLEMS ARE POSED WHICH REQUIRE MORE COMPLICATED TREATMENT.

WORKED EXAMPLE [3]

A cubical item of equipment, 25 m³ free volume, is being used with a dust having a K_{st}-value of 115 bar m s⁻¹. The equipment stands 6.1 m away from an outside wall of the factory, and it is planned to use a straight vent duct to expel the vented explosion products outside this wall. What are the vent area and P_{stat} values required to limit the reduced explosion pressure to 1.5 bar a.

ANSWER

A trial and error method is required as in Worked Example [2]:
This is not a straightforward problem because the K_{st}-value lies between two graphs. This example also demonstrates how Figure 4 may be used when reduced explosion pressures are low.

First, find an answer for 100 bar m s⁻¹:—
From Appendix 1, using the graph (Figure A6) for K_{st} = 100 bar m s⁻¹ and P_{stat} = 1.1 bar a:—

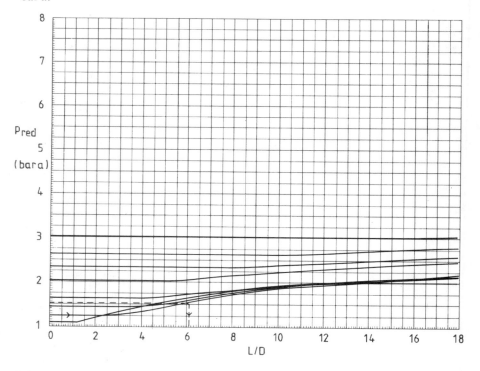

When $(P_{red})_o$ = 1.2 bar a, a duct of (L/D) ratio 6.1 is required to limit the reduced explosion pressure to 1.5 bar a.

From Figure 1 the necessary vent area in a vessel of 25 m³, with a dust of K_{st}-value 100 bar m s⁻¹, to give $(P_{red})_o$ = 1.2 bar a is 1.8 m². The vent diameter is therefore $(4 \times 1.8/3.142)^{.5}$ = 1.51 m, and thus the actual (L/D) ratio is (6.1/1.51) = 4.04.

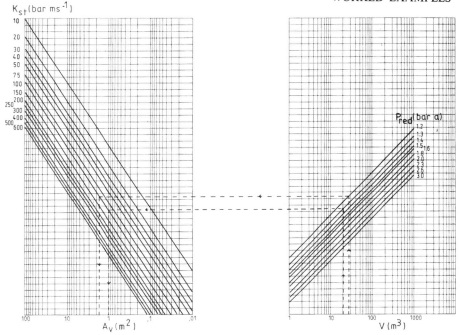

Because the actual (L/D) ratio is less than the required value, $(P_{red})_o$ can exceed 1.2 bar a.

Again using the graph from Appendix 1 for $K_{st} = 100$ bar m s^{-1} and $P_{stat} = 1.1$ bar a, when $(P_{red})_o = 1.4$ bar a, a vent duct of (L/D) ratio 5.3 will be required to lift the reduced explosion pressure to 1.5 bar a.

From Figure 1, when $K_{st} = 100$ bar m s^{-1}, the vessel volume is 25 m^3, and $(P_{red})_o = 1.4$ bar a, the vent area equals 1.0 m^2. This vent has a diameter of $(4 \times 1.0/3.142)^{.5} = 1.13$ m, and so the actual (L/D) ratio $= (6.1/1.13) = 5.4$.

A simple graphical construction provides the required $(P_{red})_o$:–

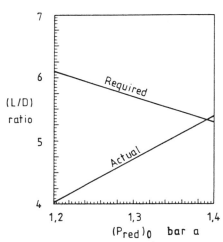

Thus, at K_{st} = 100 bar m s^{-1}, the required $(P_{red})_o$ = 1.39 Bar a, the vent area = 1.1 m^2 and P_{stat} = 1.1 bar a.

To complete the problem; the above calculations are repeated for K_{st} = 150 bar m s^{-1}.

From Appendix 1, using the graph (Figure A7) for K_{st} = 150 bar m s^{-1} and P_{stat} = 1.1 bar a:–

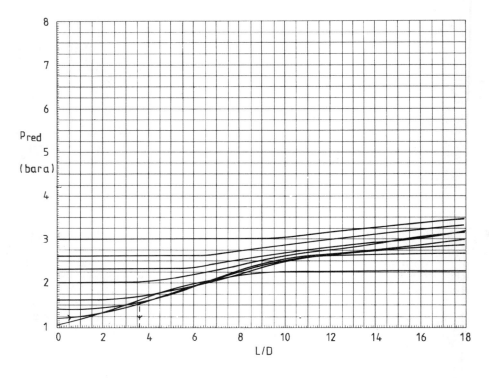

When $(P_{red})_o$ = 1.2 bar a, a duct of (L/D) ratio = 3.6 is required to give a reduced explosion pressure of 1.5 bar a.

From Figure 1, the necessary vent area in a vessel of 25 m^3, with a dust of K_{st}-value = 150 bar m s^{-1}, to give a $(P_{red})_o$ = 1.2 bar a is 2.1 m^2.

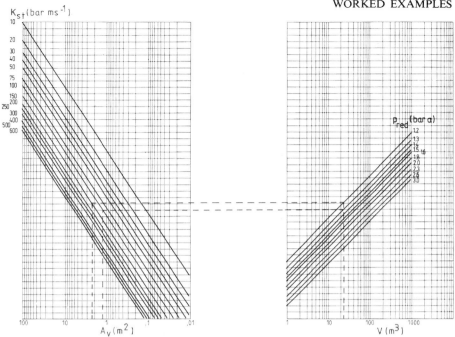

The vent diameter is therefore 1.64 m, and so the actual L/D ratio = $(6.1/1.64)$ = 3.72. Because the actual (L/D) ratio of 3.72 is greater than the required value of 3.6, $(P_{red})_o$ must be less than 1.2 bar a.

From the graph in Appendix 1 for K_{st} = 150 bar m s^{-1} and P_{stat} = 1.1 bar, when $(P_{red})_o$ = 1.05 bar, a vent duct of (L/D) ratio = 3.0 is required to raise the reduced explosion pressure to 1.5 bar a.

Because, $(P_{red})_o < 1.2$ bar a, the vent area must be obtained from Figure 4:–

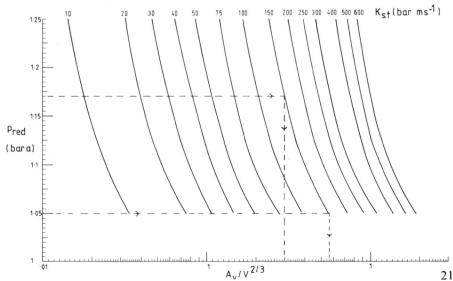

When $(P_{red})_o = 1.05$ bar a and $K_{st} = 150$ bar m s^{-1}, $(A_v/V^{2/3}) = .56$.

Thus the vent area = $(25^{.67} \times .56) = 4.84$ m^2. The vent diameter = $(4 \times 4.84/3.142)^{.5} = 2.48$ m and thus the actual (L/D) ratio is $(6.1/2.48) = 2.46$.

A simple graphical construction gives the required $(P_{red})_o$ at $K_{st} = 150$ bar m s^{-1}.

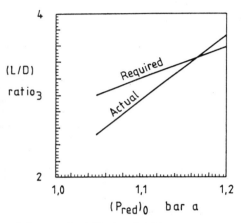

At $K_{st} = 150$ bar m s^{-1} the required $(P_{red})_o = 1.17$ bar a.

The required vent area is obtained from Figure 4 as $(25^{.67} \times .3) = 2.6$ m^2, and P_{stat} would have the value 1.08 bar a, using the rule of halving the reduced explosion overpressure when $(P_{red})_o < 1.2$ bar a.

As a check on the result, the actual (L/D) ratio of the vent duct once the vent area is known can be compared to the required (L/D) ratio from the graph for $K_{st} = 150$ bar m s^{-1} when $(P_{red})_o$ is known.

A simple graphical construction is used to find the necessary vent area at a K_{st} of 115 bar m s^{-1}.

Plot $(P_{red})_o$ against K_{st}-value:—

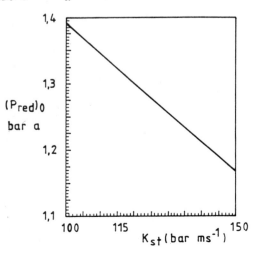

WORKED EXAMPLES

Thus at $K_{st} = 115$ bar m s^{-1}, the required $(P_{red})_o = 1.32$ bar a. The necessary vent area is found, from Figure 1, and equals 1.2 m^2. The value of $P_{stat} = 1.1$ bar a. (Had the required $(P_{red})_o$ fallen below 1.2 bar a, the necessary vent area would have been found from Figure 4, and the P_{stat} calculated by halving the reduced explosion overpressure i.e

$$P_{stat} = \left(1 + \frac{(P_{red}-1)}{2}\right) \text{ bar a}).$$

Thus the answer is:–
Vent Area = 1.2 m^2 and P $_{stat}$ = 1.1 bar a.

WORKED EXAMPLE [4]

A vessel of 100 m³ capacity handling a dust with a K_{st} value of 125 bar m s⁻¹ is fitted with a vent of area 3.9 m², which when closed with a vent panel having P_{stat} = 1.5 bar a, gives a reduced explosion pressure $(P_{red})_o$ of 1.8 bar a, according to the K_{st}-Nomographs. Determine the reduced explosion pressure when a vent duct 9 m long and containing a single 90° bend is fitted outside the vent opening. Estimate the new vent area and new P_{stat} required to ensure that the reduced explosion pressure does not exceed 1.8 bar a.

ANSWER

The vent diameter, $D, = (4 \times 3.9/3.142)^{1/2} = 2.23$ m
Assume L is measured by the methods given in Figure 5, ∴ $(L/D) = 4.04$
From Appendix 3, using the graph (Figure A108) for $K_{st} = 100$ bar m s⁻¹ and $P_{stat} = 1.5$ bar a:

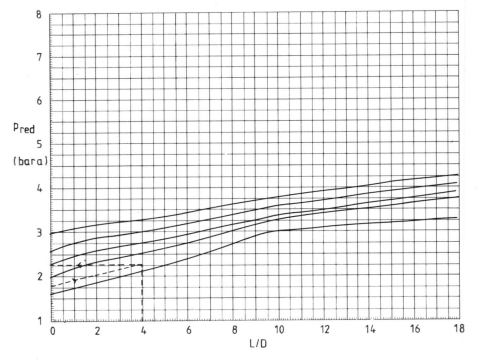

When $(P_{red})_o = 1.8$ bar a and $L/D = 4.04$, the reduced explosion pressure = 2.30 bar a.
From Appendix 3, using the graph (Figure A109) for $K_{st} = 150$ bar m s⁻¹ and $P_{stat} = 1.5$ bar a:

WORKED EXAMPLES

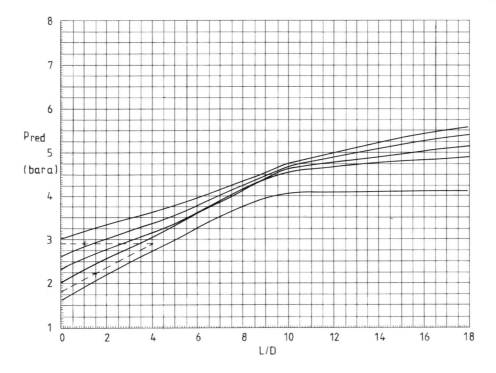

When $(P_{red})_o = 1.8$ bar a and $L/D = 4.04$, the reduced explosion pressure = 2.9 bar a. And by linear interpolation, at $K_{st} = 125$ bar m s^{-1} the new reduced explosion pressure is
$P_{red} = 2.6$ *bar a*.

Finding a new vent area to return the reduced explosion pressure to 1.8 bar a requires the trial and error method, but simply increasing the vent area is not effective. The vent area must be increased in tandem with a decrease in the vent opening pressure, P_{stat}.

In this example, an exact vent area is not required, only one which without being excessively overestimated will reduce the reduced explosion pressure to 1.8 bar a.

From Appendix 3, using the graph (Figure A95) for $K_{st} = 100$ bar m s^{-1} and $P_{stat} = 1.2$ bar a.

25

A GUIDE TO DUST EXPLOSION, PART 3

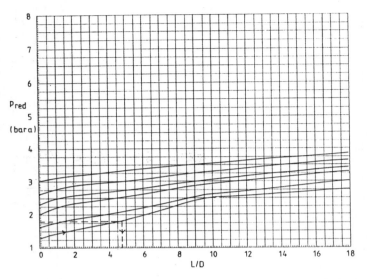

When $(P_{red})_o = 1.3$ bar a, a vent duct of $(L/D) = 4.8$ is required to give an explosion pressure of 1.8 bar a.

From the K_{st}–Nomograph in Figure 2, the vent area necessary to give a $(P_{red})_o = 1.3$ bar a in a volume of 100 m³ for $K_{st} = 100$ bar m s⁻¹ is 3.1 m².

The vent diameter, D, = $(4 \times 3.1/3.142)^{1/2} = 1.99$ m. Thus the actual (L/D) ratio = $9/1.99 = 4.52$.

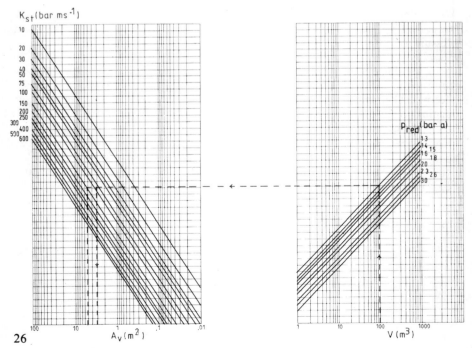

From Appendix 3, using the graph (Figure A96) for K_{st} = 150 bar m s^{-1} and P_{stat} = 1.2 bar a:

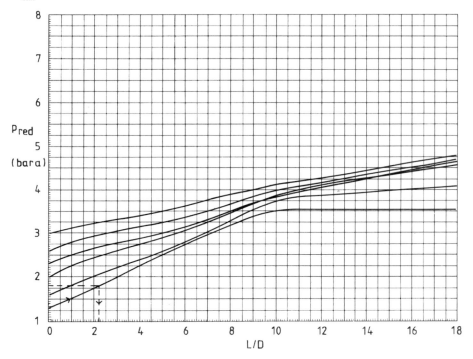

When $(P_{red})_o$ = 1.3 bar a, a vent duct of (L/D) = 2.2 is required to give an explosion pressure of 1.8 bar a.

From the K_{st}–Nomograph in Figure 2, the vent area for $(P_{red})_o$ = 1.3 bar a with K_{st} = 150 bar m s^{-1} is 5 ,^{2}. The vent diameter, D, = $4 \times 5/3.142)^{1/2}$ = 2.52. Thus the actual (L/D) = 9/2.52 = 3.57.

A simple graph demonstrates that at a K_{st} of 125 bar m s^{-1} the actual (L/D) ratio of the vent duct exceeds the required (L/D) ratio to limit the reduced pressure to 1.8 bar a.

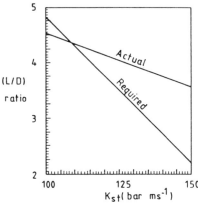

Thus, at K_{st} = 125 bar m s^{-1}, the actual (L/D) ratio exceeds the required value when $(P_{red})_o$ = 1.3 bar a and P_{stat} = 1.2 bar a. So P_{stat} needs to be reduced further.

From Appendix 3, using the graph (Figure A82) for K_{st} = 100 bar m s^{-1} and P_{stat} = 1.1 bar a.

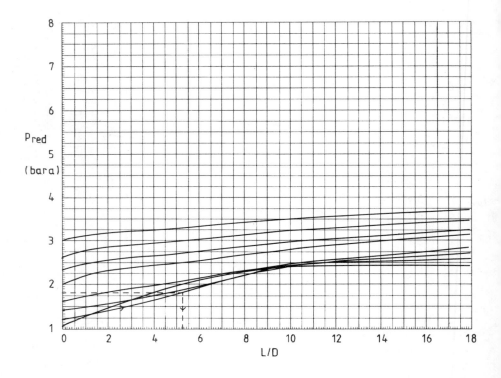

When $(P_{red})_o$ = 1.2 bar a, a vent duct of (L/D) = 5.3 is required to give a reduced explosion pressure of 1.8 bar.

WORKED EXAMPLES

From the K_{st} Nomograph in Figure 1, the vent area to give $(P_{red})_o = 1.2$ bar a in a volume of 100 m^3 with $K_{st} = 100$ bar m s^{-1} is 4 m^2. The vent diameter, D, $= (4 \times 4/3.142)^{1/2} = 2.26$. Thus the actual (L/D) ratio $= 9/2.26 = 3.98$.

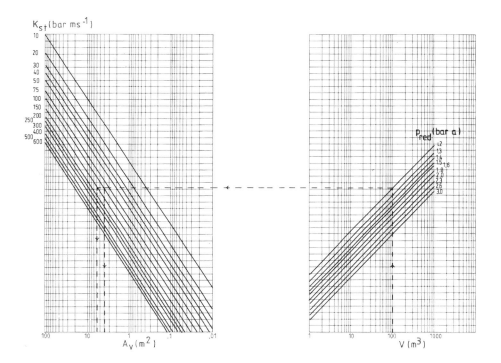

From Appendix 3, using the graph (Figure A83) for K_{st} = 150 bar m s^{-1} and P_{stat} = 1.1 bar a:

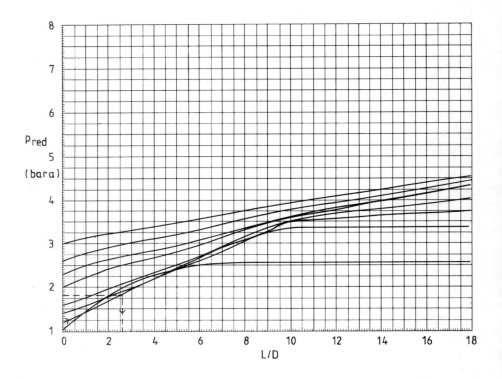

When $(P_{red})_o$ = 1.2 bar a, a vent duct of (L/D) = 2.6 is required to give a reduced explosion pressure of 1.8 bar a.

From the K_{st}-Nomograph in Figure 1, the vent area for $(P_{red})_o$ = 1.2 bar m s^{-1} is 6 m^2. The vent diameter, D, = $(4 \times 6/3.142)^{1/2}$ = 2.76. The actual (L/D) ratio = 9/2.76 = 3.26

Again, using a simple graphical construction, the actual (L/D) of the vent duct is less than the (L/D) ratio required to limit the reduced explosion pressure to 1.8 bar a at $K_{st} = 125$ bar m s^{-1}.

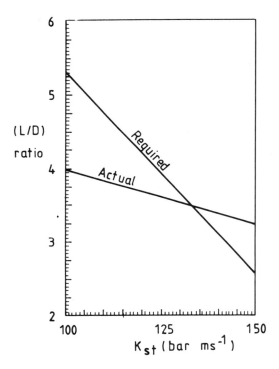

Thus, at $K_{st} = 125$ bar m s^{-1}, the actual (L/D) ratio is less than the required value when $(P_{red})_o$ is 1.2 bar a and $P_{stat} = 1.1$ bar a. The vent area can be obtained from the K_{st}-Nomograph in Figure 1, and is 5 m².

Therefore a vent with an area of 5 m², with a vent cover of $P_{stat} = 1.1$ bar a will ensure, in the conditions described that the reduced explosion pressure will be less than 1.8 bar a.

If the value of vent area that limits the reduced pressure exactly to 1.8 bar a is required, then the procedures described in WORKED EXAMPLES [3] and [5] will need to be applied at $P_{stat} = 1.1$ bar a.

A GUIDE TO DUST EXPLOSION, PART 3

WORKED EXAMPLE [5]

A vessel of free volume 45 m³ is used to handle dust with a K_{st}-value of 100 bar m s⁻¹. The vent is covered by a panel with a bursting pressure of 1.2 bar a. The vent is 8 m from an external wall and a straight duct will be used to vent any explosion safely. Estimate the vent area required to limit the reduced explosion pressure to 1.7 bar a.

ANSWER

From Appendix 1, using the graph (Figure A19) for K_{st} = 100 bar m s⁻¹ and P_{stat} = 1.2 bar a:–

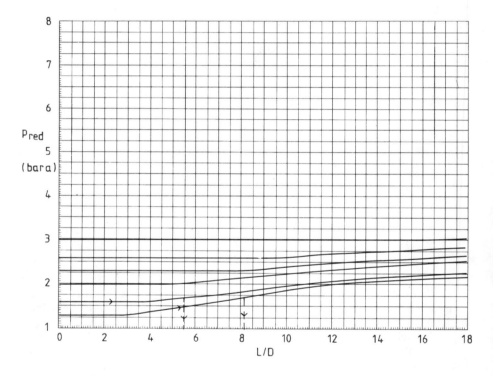

When $(P_{red})_o$ = 1.3 bar a, a vent duct of (L/D) = 8.2 is required to give an explosion pressure of 1.7 bar a.

From the K_{st}-Nomograph in Figure 2, when $(P_{red})_o$ = 1.3 bar a and K_{st} = 100 bar m s⁻¹ the required vent area in a volume of 45 m³ is 2 m². This vent diameter = $(4 \times 2/3.142)^{1/2}$ = 1.59 m and this gives an actual (L/D) ratio of $(8/1.59)$ = 5.03, which is less than the required (L/D) ratio.

WORKED EXAMPLES

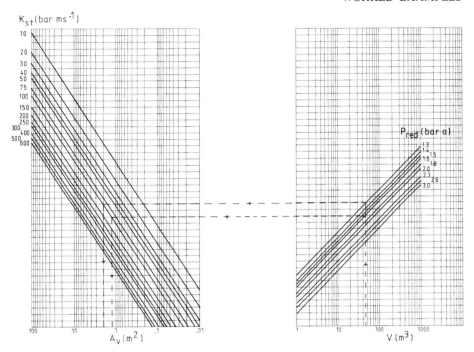

When $(P_{red})_o = 1.6$ bar a, a vent duct of $(L/D) = 5.5$ is required to give an explosion pressure of 1.7 bar a.

From the K_{st}-Nomograph in Figure 2, when $(P_{red})_o = 1.6$ bar a and $K_{st} = 100$ bar m s^{-1} the required vent area in a volume of 45 m^2 is 1.1 m^2. This vent diameter is $(4 \times 1.1/3.142)^{1/2} = 1.18$ m. This gives an actual (L/D) ratio of $(8/1.18) = 6.78$, which is greater that the rquired (L/D) ratio.

Using a simple graphical construction:–

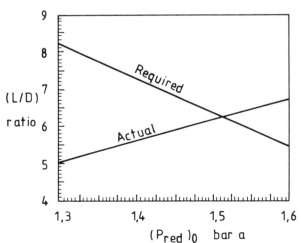

At $(P_{red})_o = 1.52$ bar a, the actual (L/D) equals the required (L/D) value, and from the K_{st}-Nomograph in Figure 2, the required vent area equals 1.3 m².

$$A_V = 1.3 \text{ m}^2$$

When the dust has a K_{st}-value which lies between graphs, follow the above procedure using the graphs for the K_{st}-values directly below and above the actual value, and interpolate to obtain the vent area at the actual K_{st}-value.

3. OTHER DESIGN INFORMATION

3.1 POSITION OF THE BEND ALONG THE DUCT
The experiments on which this design is based were mostly done with a bend one metre from the duct exit. The bend must not be close to the vent opening itself. It is recommended that the bend be at least 2 m from the vent opening.

3.2 EFFECT OF MULTIPLE BENDS
It is recommended that single bends only be incorporated in a vent duct. Experiments show that the effect of two bends on the reduced explosion pressure is greater than suggested by assuming that the contributions are additive.

3.3 STRENGTH OF THE DUCTS
Explosion pressures in the vent duct can be as high as those measured in the explosion vessel. The ducting should thus be made strong enough to withstand the expected explosion pressures. Pressures in the vent duct can sometimes be higher than those measured in the explosion vessel. These are however, short lived pressure pulses.

3.4 GRADUAL BENDS
Sharp bends have been used in this guide because they represent the worst case. If a gradual bend is used the pressure will be lower than for a sharp bend, all other things being equal. It is recommended that the outer curve of the duct be used when calculating the L/D ratio of a gradual bend, (See Figure 5) and Appendix 1 used. When the ratio (radius of curvature of the bend/duct diameter) is less than 2, however, the length L should be measured in the same way but the bend considered as sharp, and Appendix 2 and 3 used.

3.5 REACTION FORCES
Both explosion vessels and ducting should be securely fixed to withstand reaction forces. If the vent duct contains a bend, the lateral reaction forces should also be catered for.

3.6 PROTECTING THE VENT DUCT BY AN END GRATING
A grating over the end of a vent duct does not have a noticeable effect on the reduced explosion pressure, providing the mesh is not too small.

REFERENCES

1. Schofield, C., 1985, *Guide to Dust Explosion Prevention and Protection, Part 1 – Venting.* (The Institution of Chemical Engineers, Rugby, England).
2. Bartknecht, W., 1981, *Explosions, Cause, Prevention, Protection.* (Springer Verlag, Berlin, Heidelberg, New York).
3. Heinrich, H.J., 1966, *Chemie Ing Techn* 38: 1125.
4. VDI 3673, 1983, *Pressure Release of Dust Explosions.* (Verein Deutscher Ingeneuire).

APPENDIX 1

ESTIMATES OF REDUCED EXPLOSION PRESSURES FOR STRAIGHT VENT DUCTS

NOTE: All pressures are given in bar a.

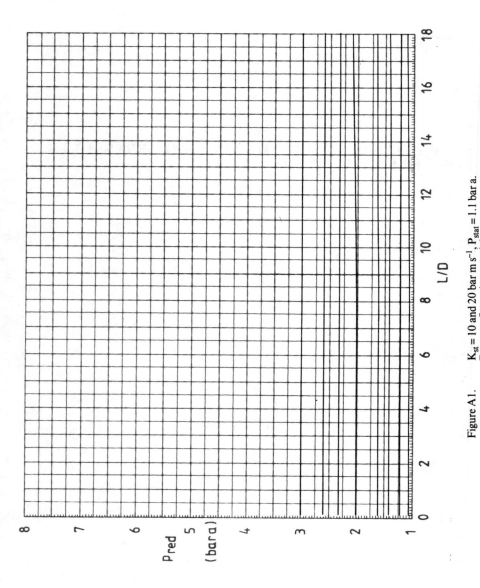

Figure A1. $K_{st} = 10$ and $20 \, \text{bar m s}^{-1}$; $P_{stat} = 1.1 \, \text{bar a}$. Duct configuration: straight.

APPENDICES

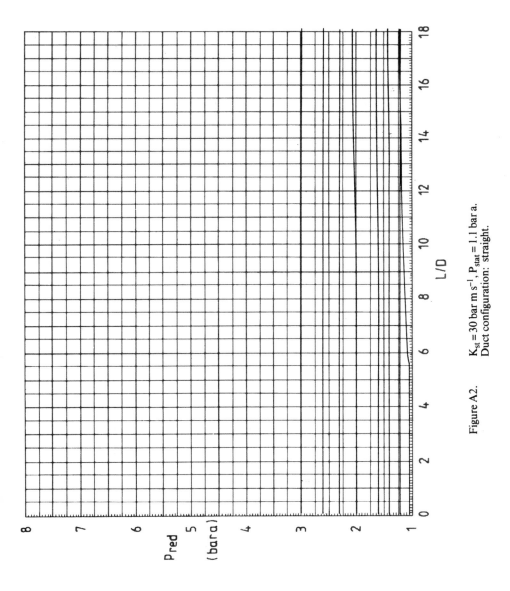

Figure A2. $K_{st} = 30$ bar m s^{-1}, $P_{stat} = 1.1$ bar a. Duct configuration: straight.

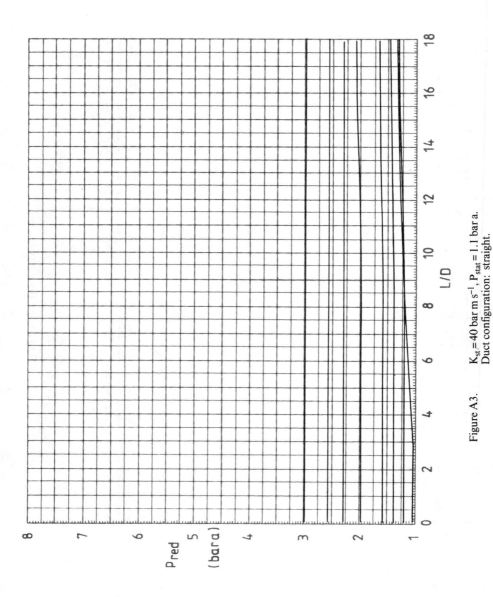

Figure A3. $K_{st} = 40$ bar m s^{-1}, $P_{stat} = 1.1$ bar a. Duct configuration: straight.

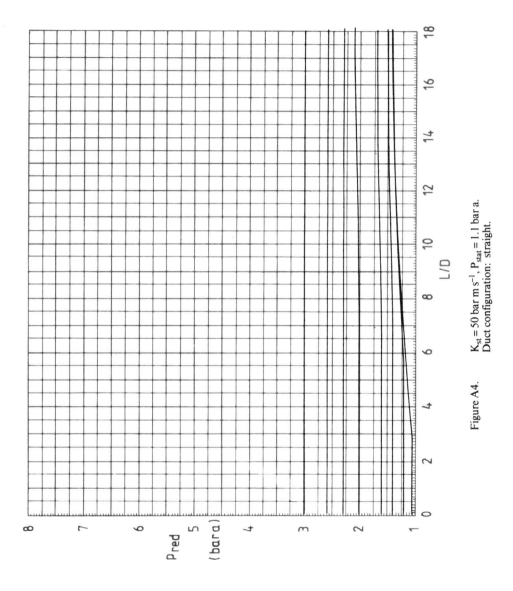

Figure A4. $K_{st} = 50$ bar m s^{-1}, $P_{stat} = 1.1$ bar a. Duct configuration: straight.

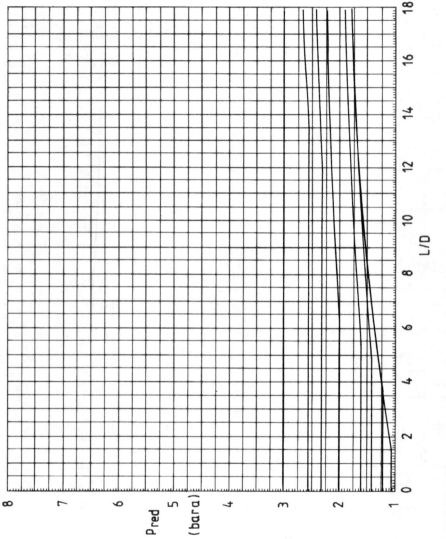

Figure A5. $K_{st} = 75$ bar m s^{-1}, $P_{stat} = 1.1$ bar a. Duct configuration: straight.

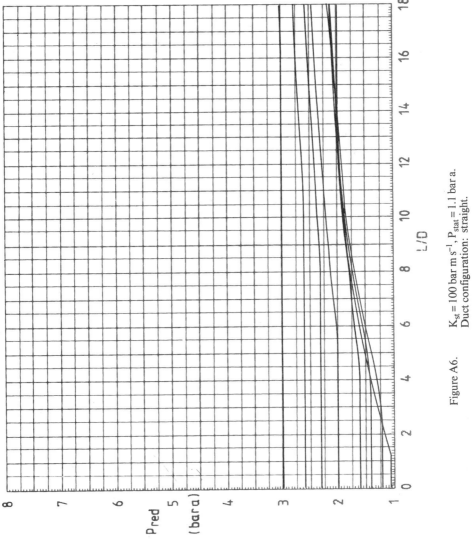

Figure A6. $K_{st} = 100$ bar m s^{-1}, $P_{stat} = 1.1$ bar a. Duct configuration: straight.

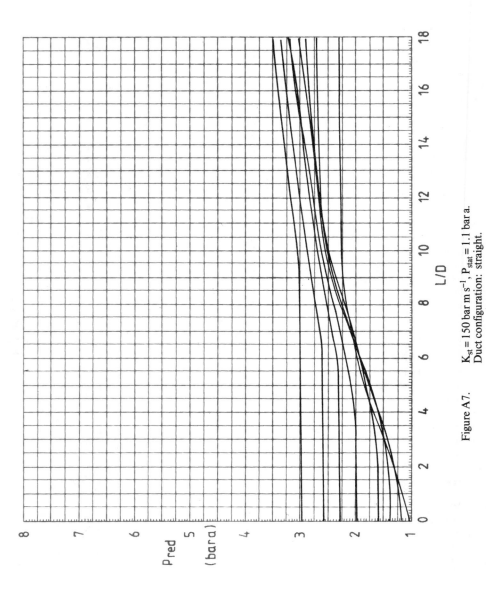

Figure A7. $K_{st} = 150$ bar m s^{-1}, $P_{stat} = 1.1$ bar a. Duct configuration: straight.

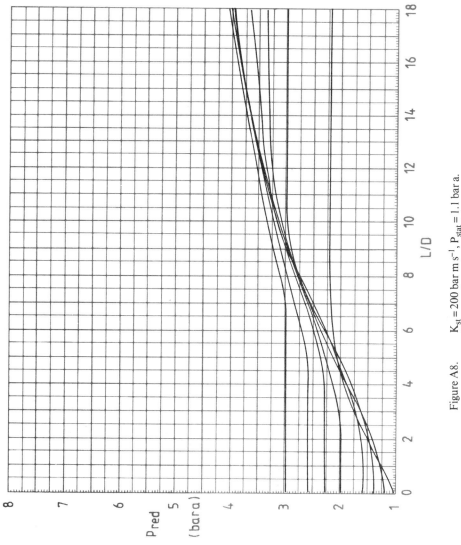

Figure A8. $K_{st} = 200$ bar m s^{-1}, $P_{stat} = 1.1$ bar a. Duct configuration: straight.

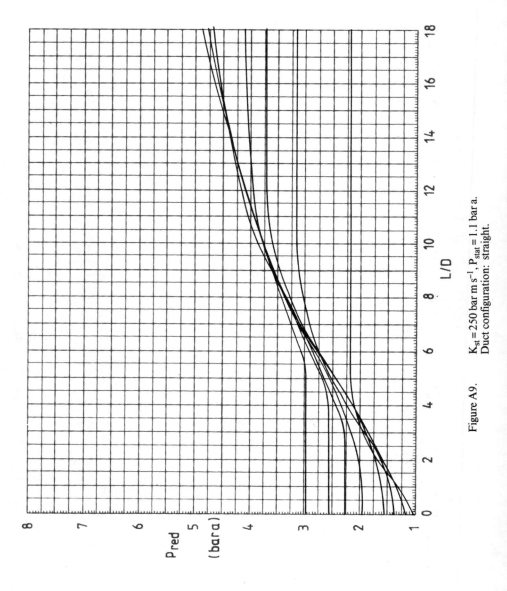

Figure A9. $K_{st} = 250$ bar m s^{-1}, $P_{stat} = 1.1$ bar a. Duct configuration: straight.

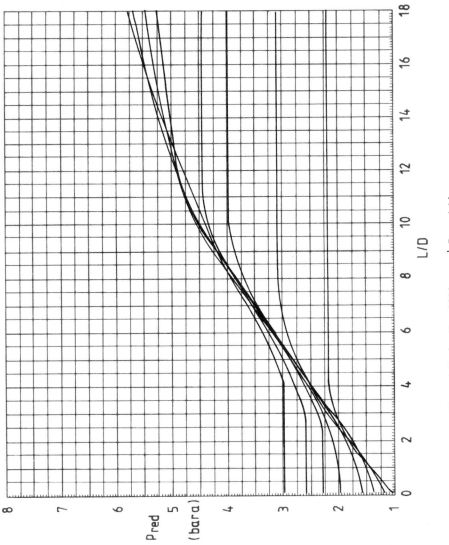

Figure A10. $K_{st} = 300$ bar m s^{-1}, $P_{stat} = 1.1$ bar a. Duct configuration: straight.

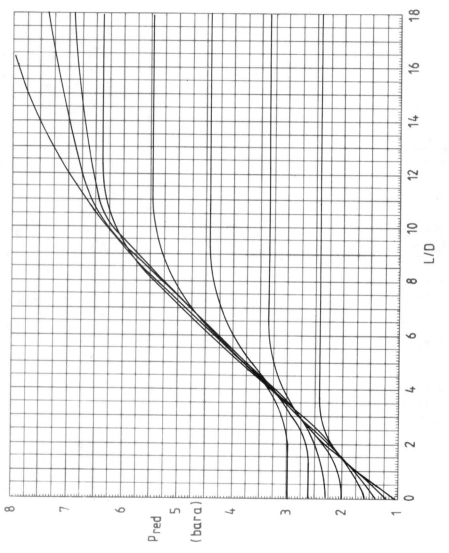

Figure A11. $K_{st} = 400$ bar m s^{-1}, $P_{stat} = 1.1$ bar a. Duct configuration: straight.

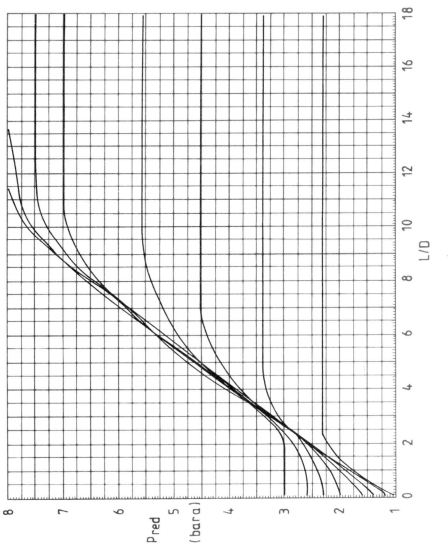

Figure A12. $K_{st} = 500$ bar m s^{-1}, $P_{stat} = 1.1$ bar a. Duct configuration: straight.

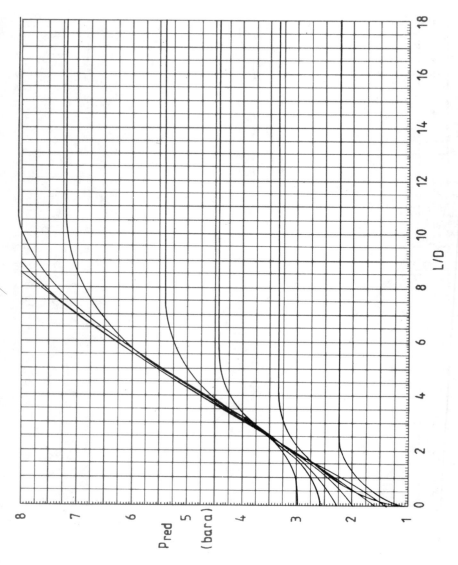

Figure A13. $K_{st} = 600$ bar m s^{-1}, $P_{stat} = 1.1$ bar a. Duct configuration: straight.

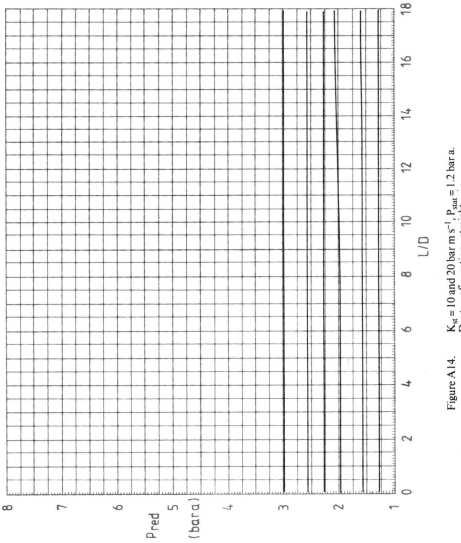

Figure A14. $K_{st} = 10$ and $20\ \text{bar m s}^{-1}$, $P_{stat} = 1.2\ \text{bar a}$. Duct configuration: straight.

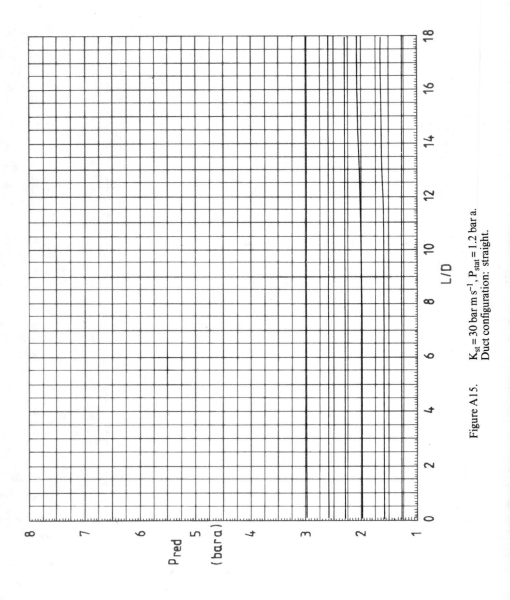

Figure A15. $K_{st} = 30$ bar m s^{-1}, $P_{stat} = 1.2$ bar a. Duct configuration: straight.

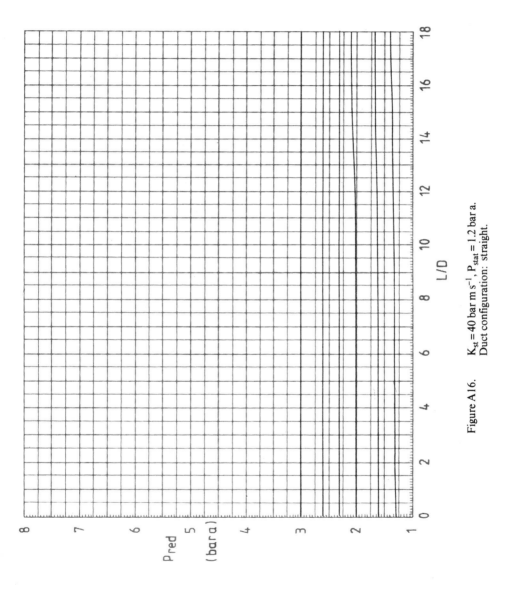

Figure A16. $K_{st} = 40$ bar m s^{-1}, $P_{stat} = 1.2$ bar a. Duct configuration: straight.

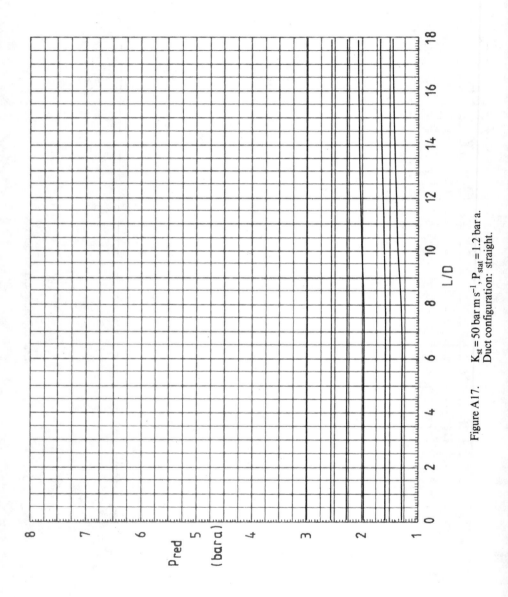

Figure A17. $K_{st} = 50$ bar m s^{-1}, $P_{stat} = 1.2$ bar a. Duct configuration: straight.

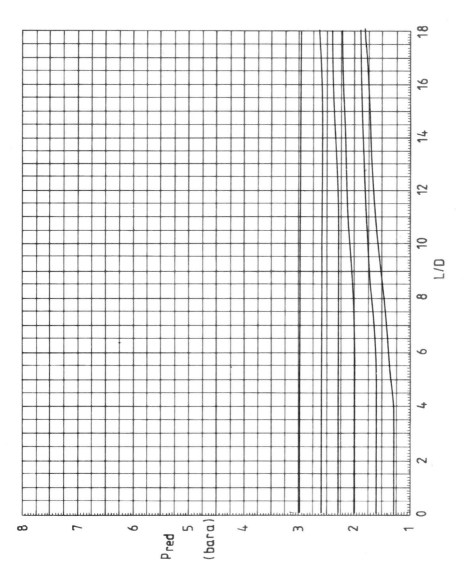

Figure A18. $K_{st} = 75$ bar m s^{-1}, $P_{stat} = 1.2$ bar a. Duct configuration: straight.

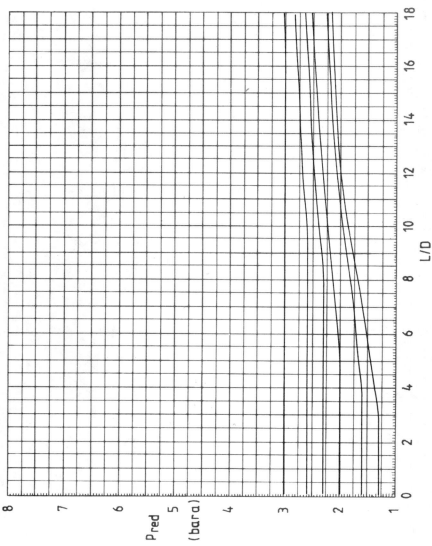

Figure A19. $K_{st} = 100$ bar m s^{-1}, $P_{stat} = 1.2$ bar a. Duct configuration: straight.

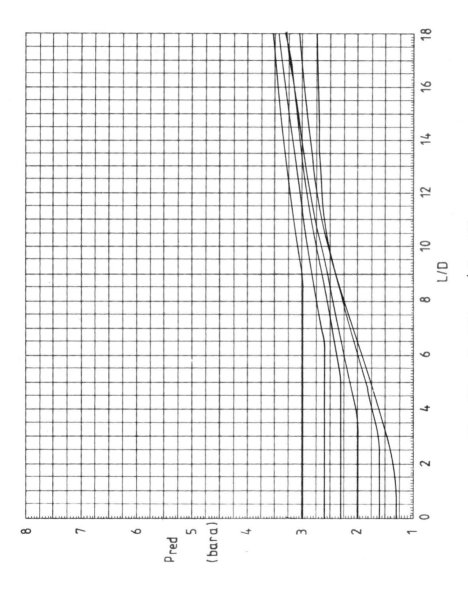

Figure A20. $K_{st} = 150$ bar m s^{-1}, $P_{stat} = 1.2$ bar a. Duct configuration: straight.

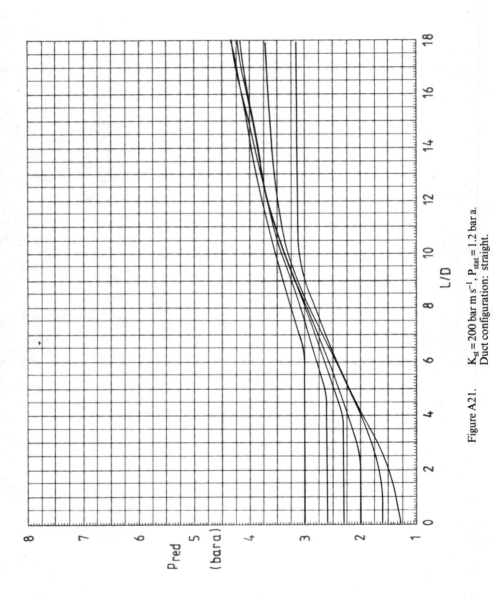

Figure A21. $K_{st} = 200$ bar m s^{-1}, $P_{stat} = 1.2$ bar a. Duct configuration: straight.

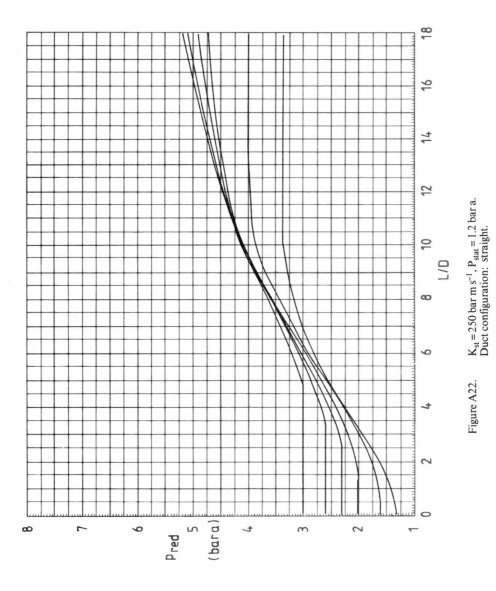

Figure A22. $K_{st} = 250$ bar m s^{-1}, $P_{stat} = 1.2$ bar a. Duct configuration: straight.

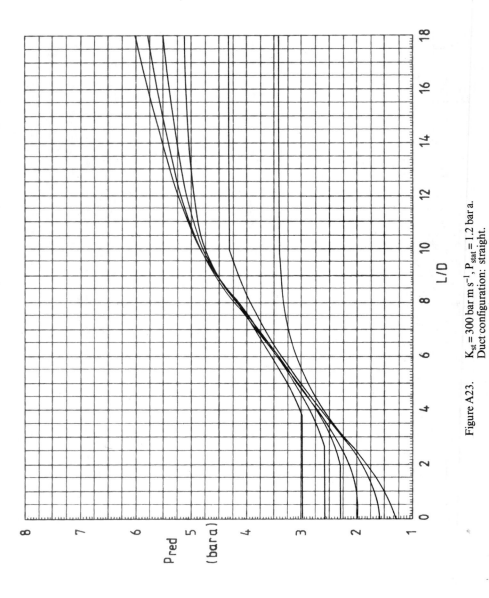

Figure A23. $K_{st} = 300$ bar m s^{-1}, $P_{stat} = 1.2$ bar a. Duct configuration: straight.

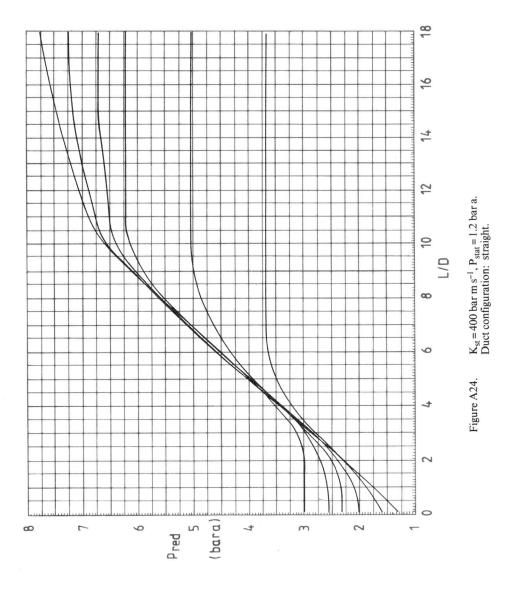

Figure A24. $K_{st} = 400$ bar m s^{-1}, $P_{stat} = 1.2$ bar a. Duct configuration: straight.

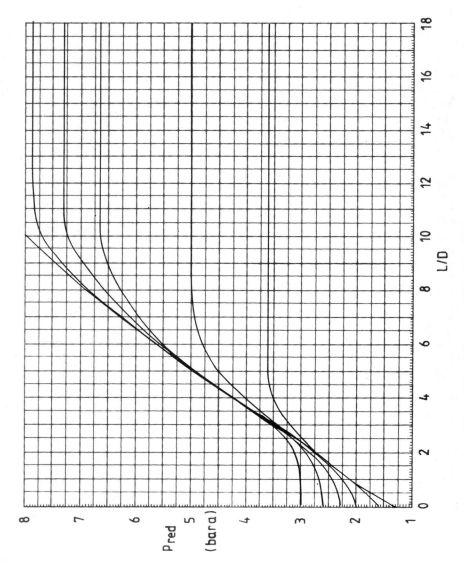

Figure A25. $K_{st} = 500$ bar m s^{-1}, $P_{stat} = 1.2$ bar a. Duct configuration: straight.

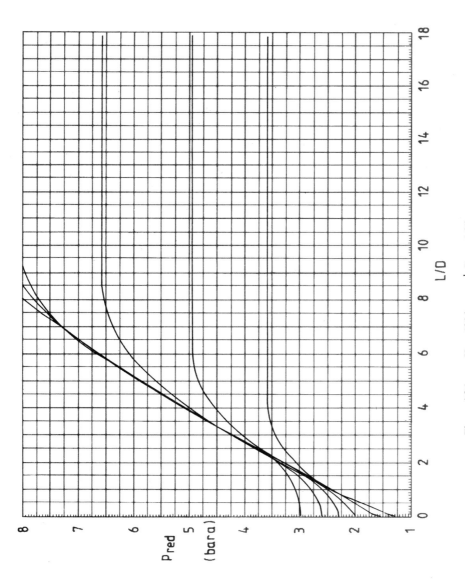

Figure A26. $K_{st} = 600$ bar m s^{-1}, $P_{stat} = 1.2$ bar a. Duct configuration: straight.

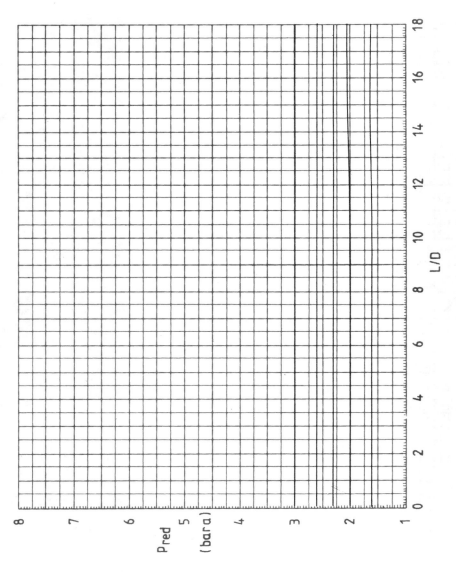

Figure A27. K_{st} = 10, 20 and 30 bar m s^{-1}, P_{stat} = 1.5 bar a. Duct configuration: straight.

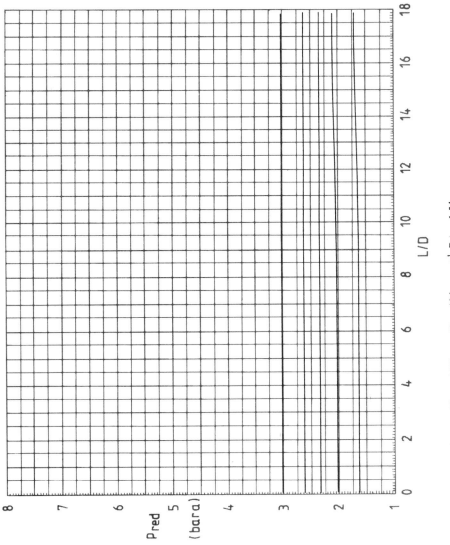

Figure A28. $K_{st} = 40$ bar m s^{-1}, $P_{stat} = 1.5$ bar a. Duct configuration: straight.

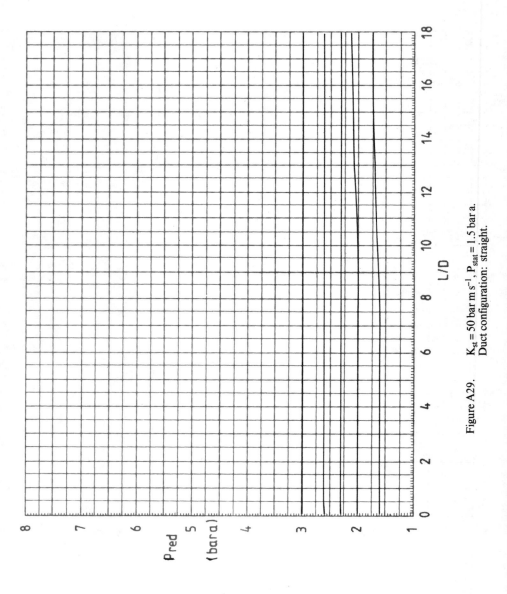

Figure A29. $K_{st} = 50$ bar m s^{-1}, $P_{stat} = 1.5$ bar a. Duct configuration: straight.

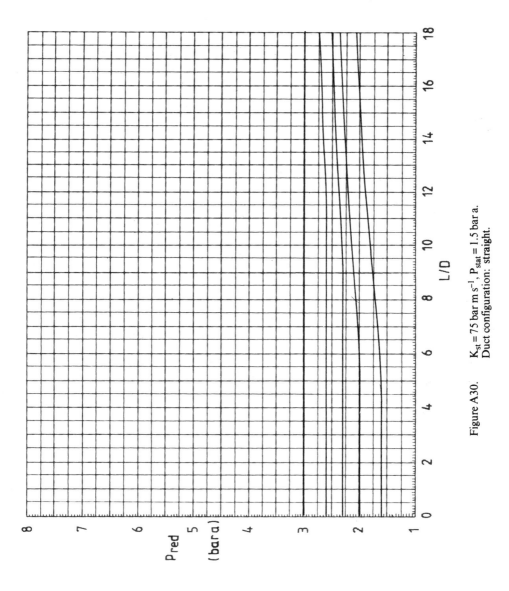

Figure A30. $K_{st} = 75$ bar m s^{-1}, $P_{stat} = 1.5$ bar a. Duct configuration: straight.

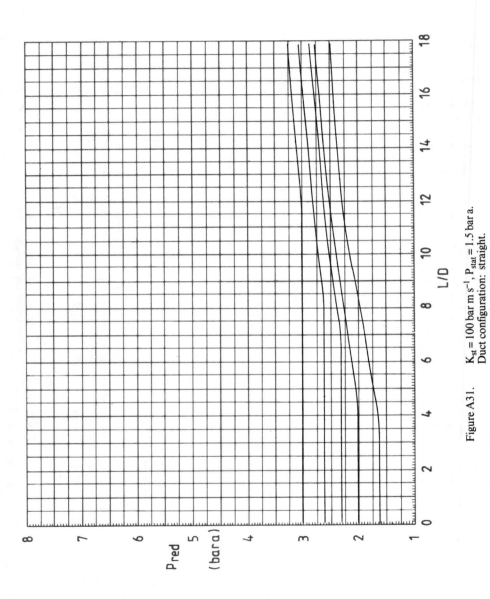

Figure A31. $K_{st} = 100$ bar m s^{-1}, $P_{stat} = 1.5$ bar a. Duct configuration: straight.

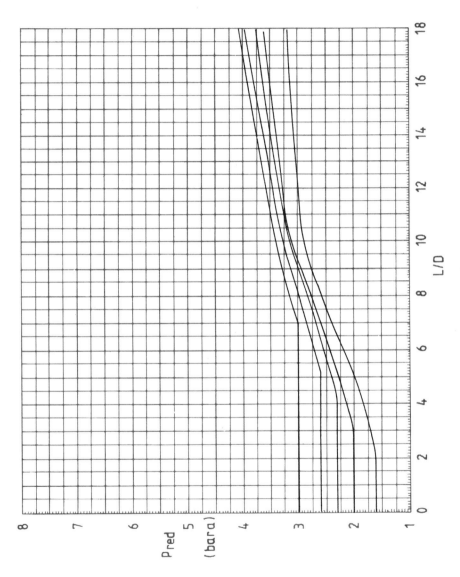

Figure A32. $K_{st} = 150$ bar m s^{-1}, $P_{stat} = 1.5$ bar a. Duct configuration: straight.

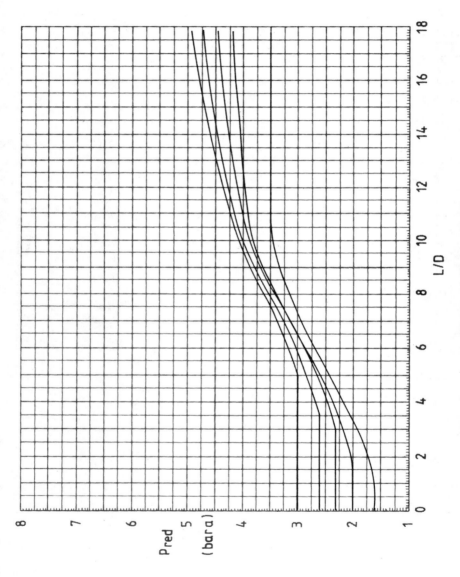

Figure A33. $K_{st} = 200$ bar m s^{-1}, $P_{stat} = 1.5$ bar a. Duct configuration: straight.

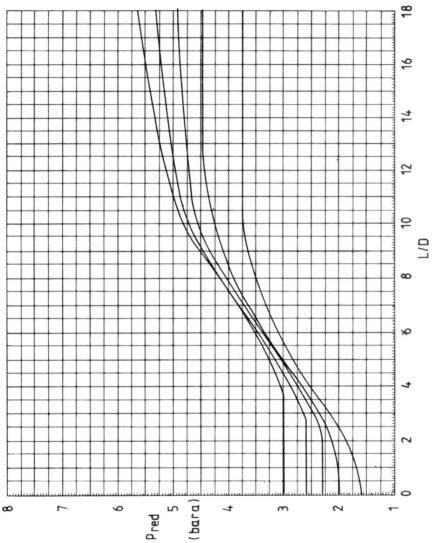

Figure A34. $K_{st} = 250$ bar m s^{-1}, $P_{stat} = 1.5$ bar a. Duct configuration: straight.

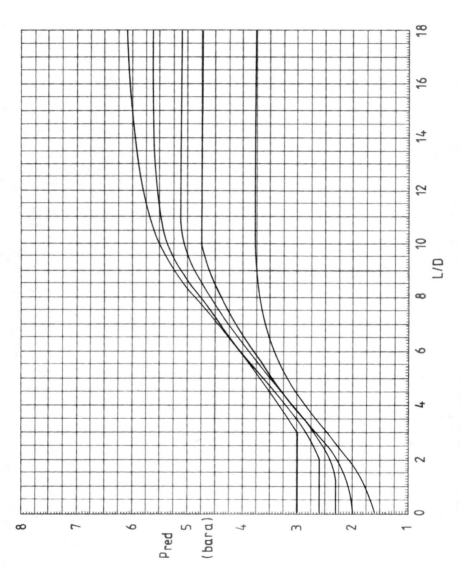

Figure A35. $K_{st} = 300 \text{ bar m s}^{-1}$, $P_{stat} = 1.5$ bar a. Duct configuration: straight.

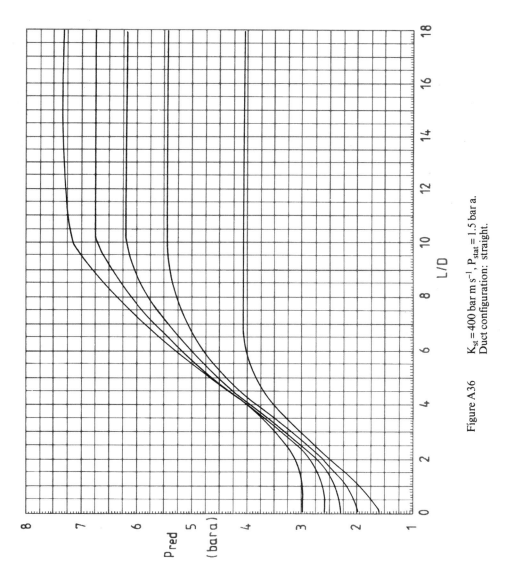

Figure A36 $K_{st} = 400$ bar m s^{-1}, $P_{stat} = 1.5$ bar a. Duct configuration: straight.

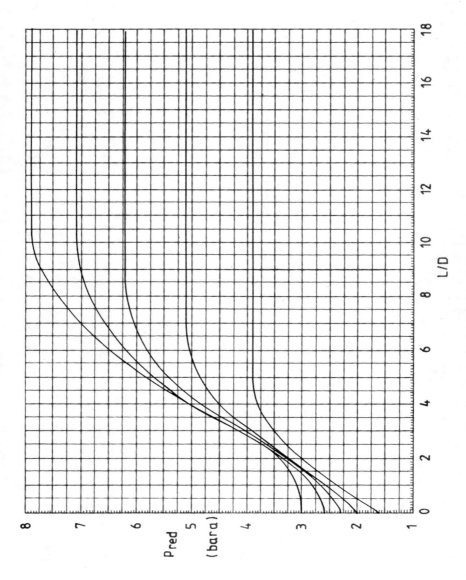

Figure A37. $K_{st} = 500$ bar m s^{-1}, $P_{stat} = 1.5$ bar a. Duct configuration: straight.

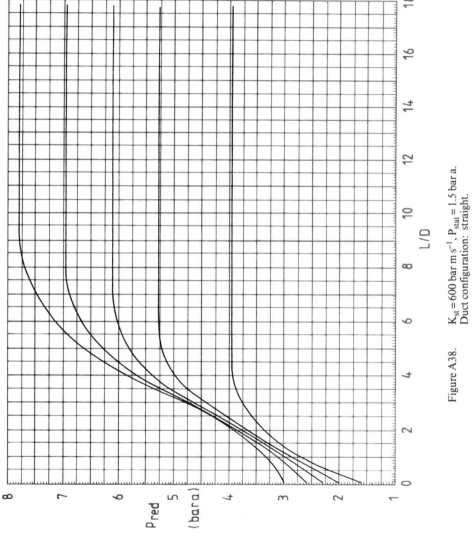

Figure A38. $K_{st} = 600$ bar m s^{-1}, $P_{stat} = 1.5$ bar a. Duct configuration: straight.

APPENDIX 2

ESTIMATES OF REDUCED EXPLOSION PRESSURES FOR VENT DUCTS WITH A SINGLE SHARP 45° BEND

NOTE: All pressures are given in bar a.

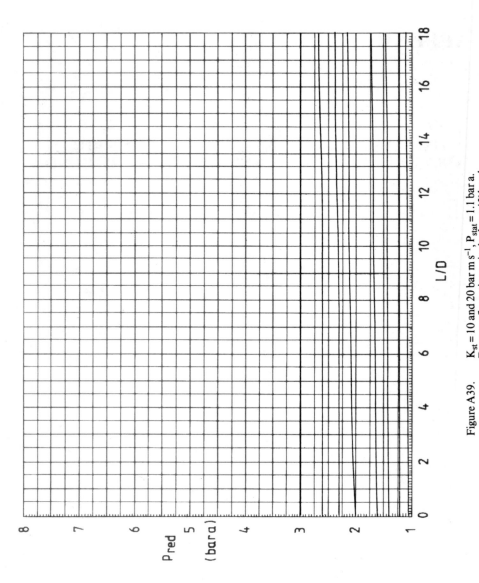

Figure A39. $K_{st} = 10$ and 20 bar m s^{-1}, $P_{stat} = 1.1$ bar a. Duct configuration: single sharp 45° bend.

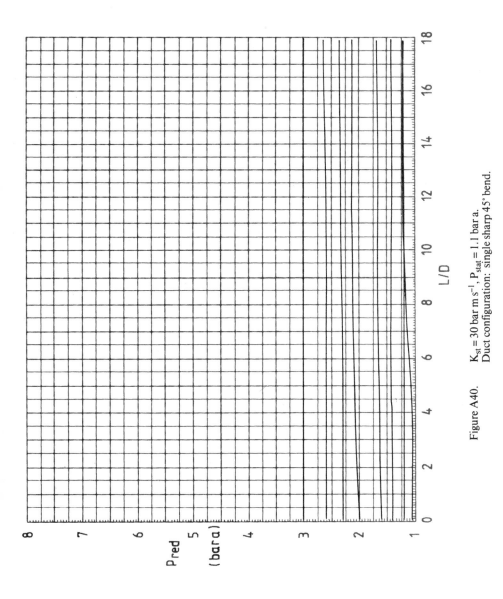

Figure A40. $K_{st} = 30$ bar m s^{-1}, $P_{stat} = 1.1$ bar a.
Duct configuration: single sharp 45° bend.

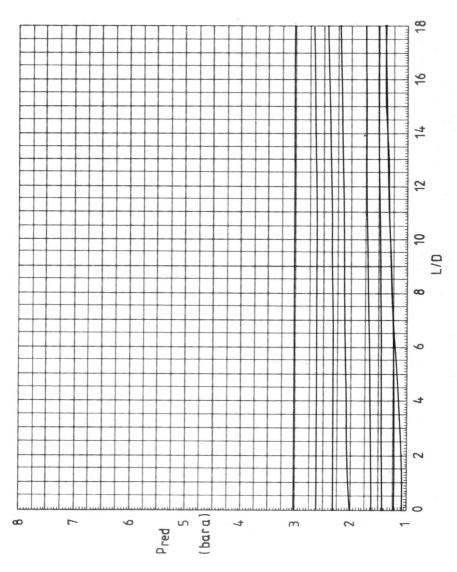

Figure A41. $K_{st} = 40$ bar m s^{-1}, $P_{stat} = 1.1$ bar a. Duct configuration: single sharp 45° bend.

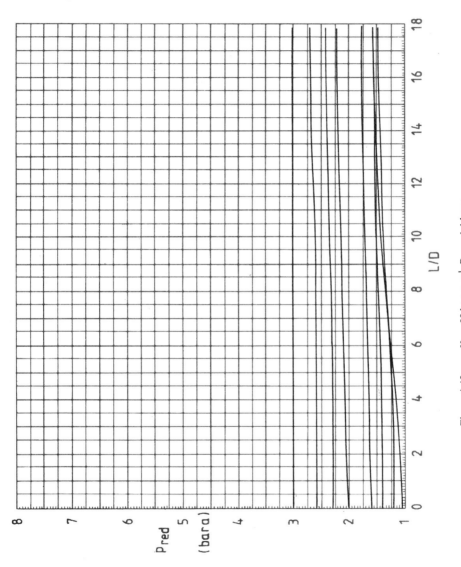

Figure A42. $K_{st} = 50$ bar m s^{-1}, $P_{stat} = 1.1$ bar a. Duct configuration: single sharp 45° bend.

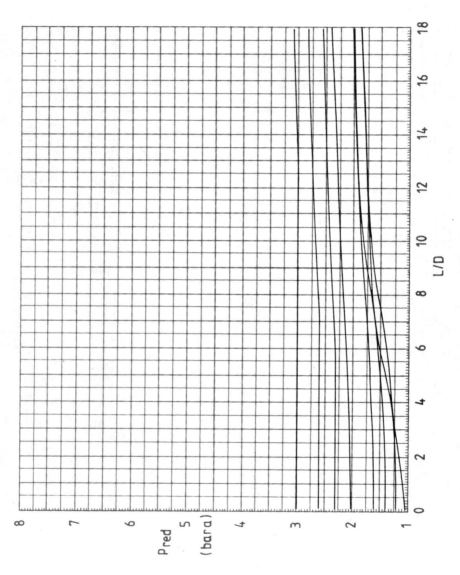

Figure A43. $K_{st} = 75$ bar m s^{-1}, $P_{stat} = 1.1$ bar a. Duct configuration: single sharp 45° bend.

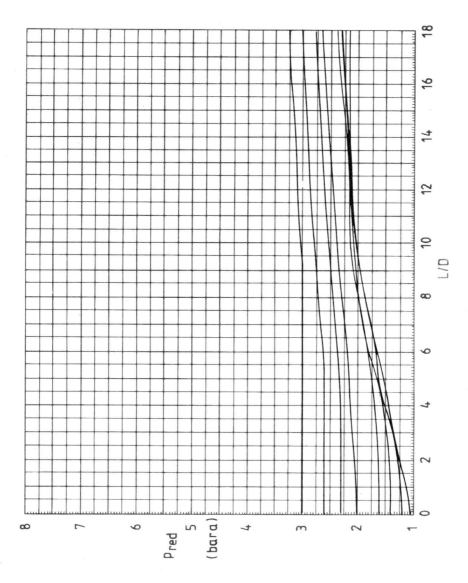

Figure A44. $K_{st} = 100$ bar m s^{-1}, $P_{stat} = 1.1$ bar a. Duct configuration: single sharp 45° bend.

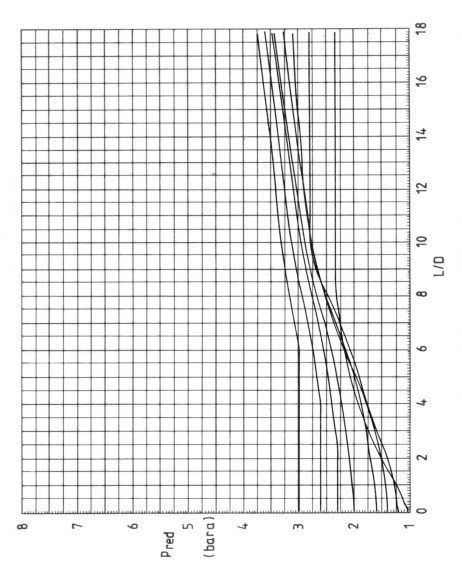

Figure A45. $K_{st} = 150$ bar m s^{-1}, $P_{stat} = 1.1$ bar a. Duct configuration: single sharp 45° bend.

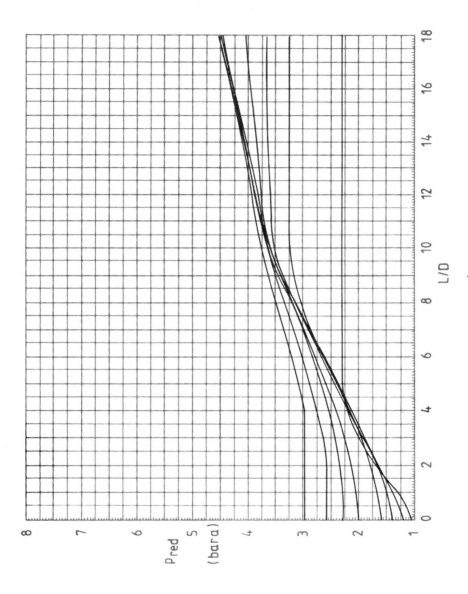

Figure A46. $K_{st} = 200$ bar m s^{-1}, $P_{stat} = 1.1$ bar a. Duct configuration: single sharp 45° bend.

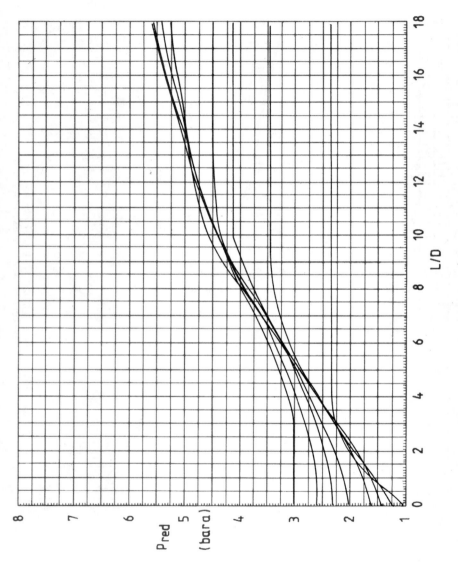

Figure A47. $K_{st} = 250$ bar m s^{-1}, $P_{stat} = 1.1$ bar a.
Duct configuration: single sharp 45° bend.

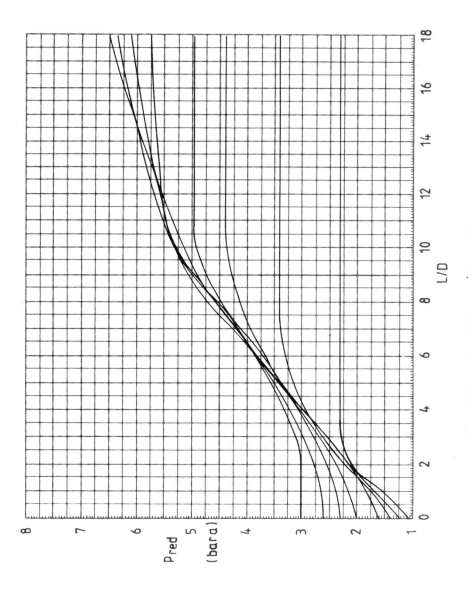

Figure A48. $K_{st} = 300$ bar m s^{-1}, $P_{stat} = 1.1$ bar a. Duct configuration: single sharp 45° bend.

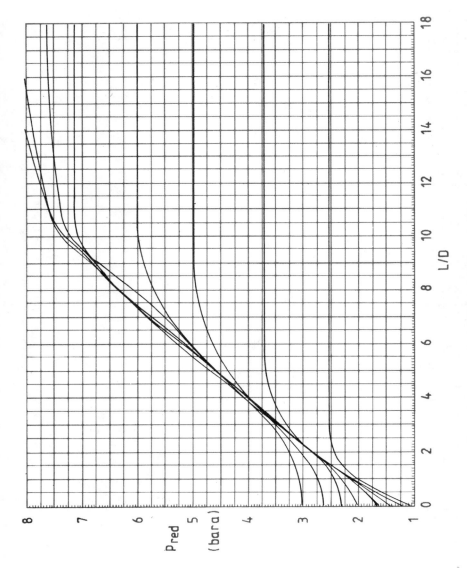

Figure A49. $K_{st} = 400$ bar m s^{-1}, $P_{stat} = 1.1$ bar a. Duct configuration: single sharp 45° bend.

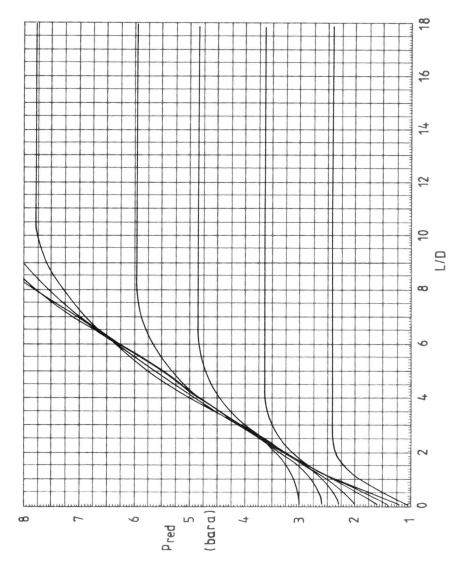

Figure A50. $K_{st} = 500$ bar m s^{-1}, $P_{stat} = 1.1$ bar a.
Duct configuration: single sharp 45° bend.

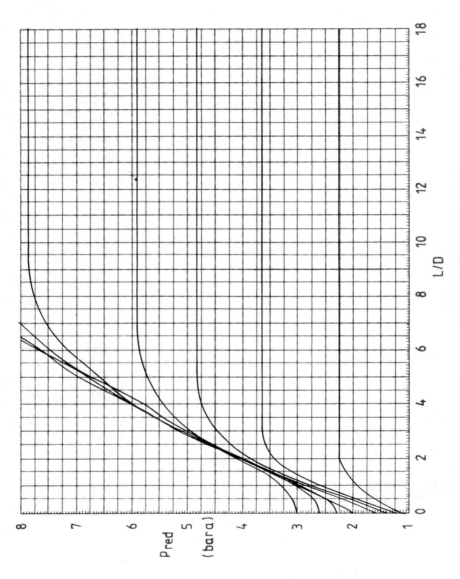

Figure A51. $K_{st} = 600$ bar m s^{-1}, $P_{stat} = 1.1$ bar a. Duct configuration: single sharp 45° bend.

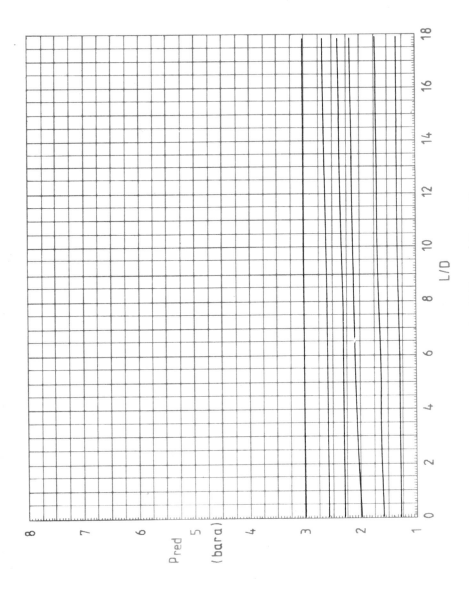

Figure A52. $K_{st} = 10$ and 20 bar m s^{-1}, $P_{stat} = 1.2$ bar a. Duct configuration: single sharp 45° bend.

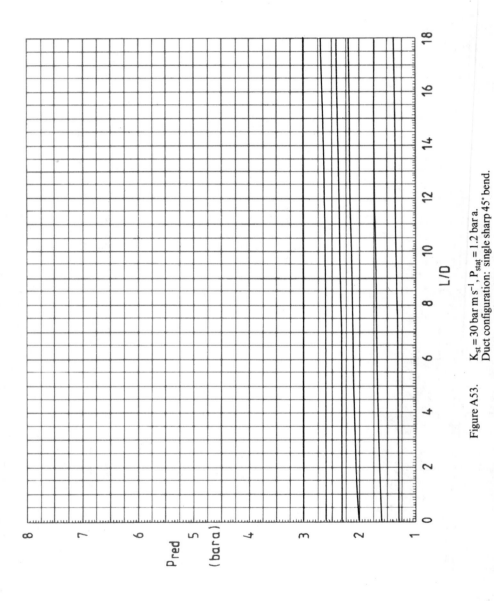

Figure A53. $K_{st} = 30$ bar m s^{-1}, $P_{stat} = 1.2$ bar a. Duct configuration: single sharp 45° bend.

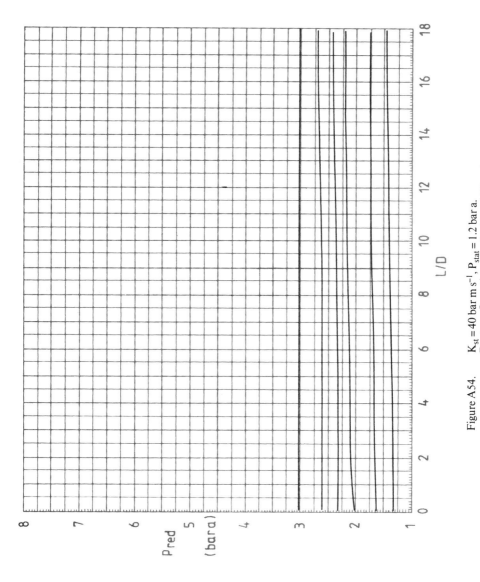

Figure A54. $K_{st} = 40$ bar m s^{-1}, $P_{stat} = 1.2$ bar a. Duct configuration: single sharp 45° bend.

A GUIDE TO DUST EXPLOSION, PART 3

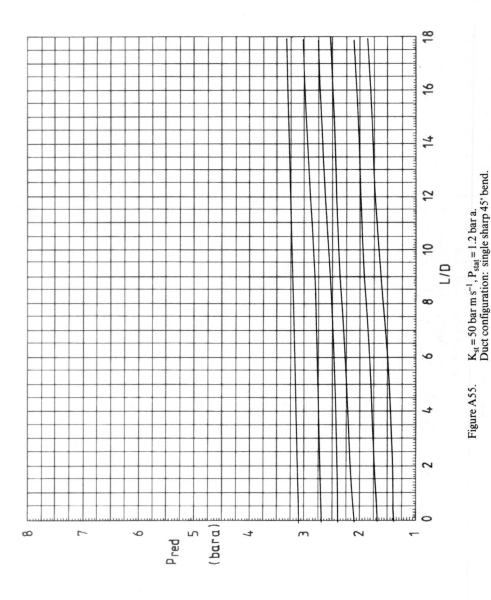

Figure A55. $K_{st} = 50$ bar m s^{-1}, $P_{stat} = 1.2$ bar a.
Duct configuration: single sharp 45° bend.

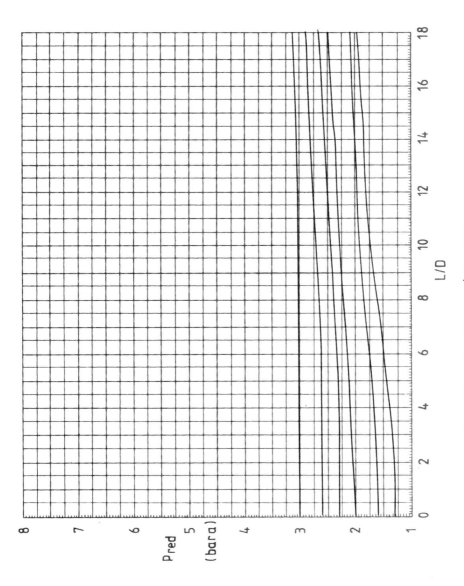

Figure A56. $K_{st} = 75$ bar m s^{-1}, $P_{stat} = 1.2$ bar a.
Duct configuration: single sharp 45° bend.

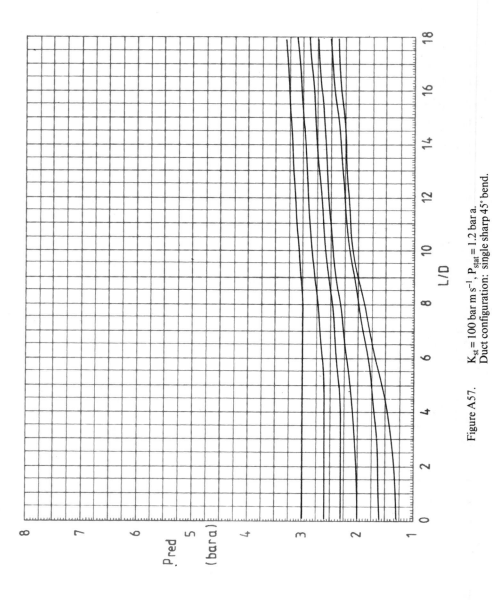

Figure A57. $K_{st} = 100$ bar m s^{-1}, $P_{stat} = 1.2$ bar a. Duct configuration: single sharp 45° bend.

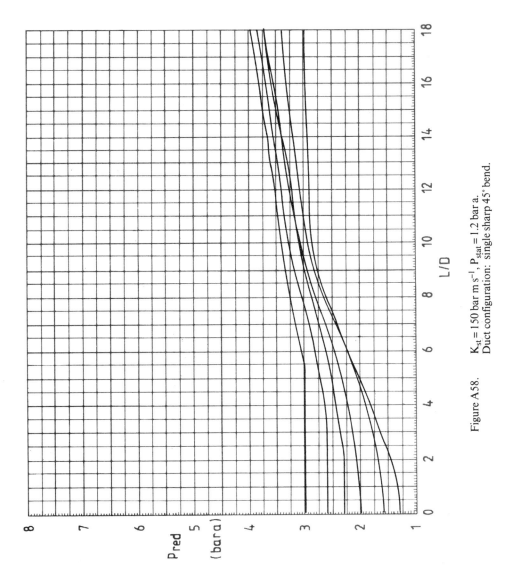

Figure A58. $K_{st} = 150$ bar m s^{-1}, $P_{stat} = 1.2$ bar a. Duct configuration: single sharp 45° bend.

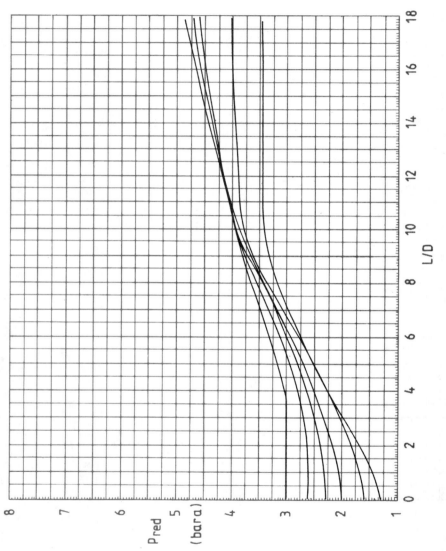

Figure A59. $K_{st} = 200$ bar m s^{-1}, $P_{stat} = 1.2$ bar a.
Duct configuration: single sharp 45° bend.

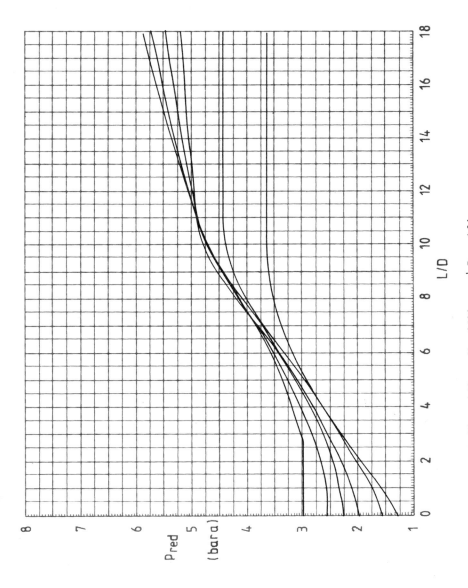

Figure A60. K_{st} = 250 bar m s^{-1}, P_{stat} = 1.2 bar a. Duct configuration: single sharp 45° bend.

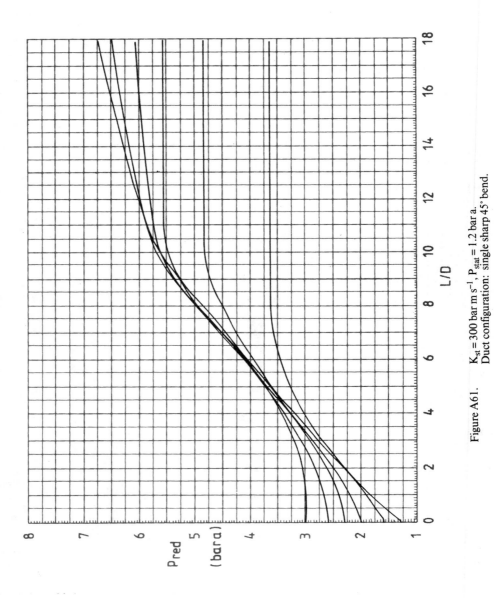

Figure A61. $K_{st} = 300$ bar m s^{-1}, $P_{stat} = 1.2$ bar a.
Duct configuration: single sharp 45° bend.

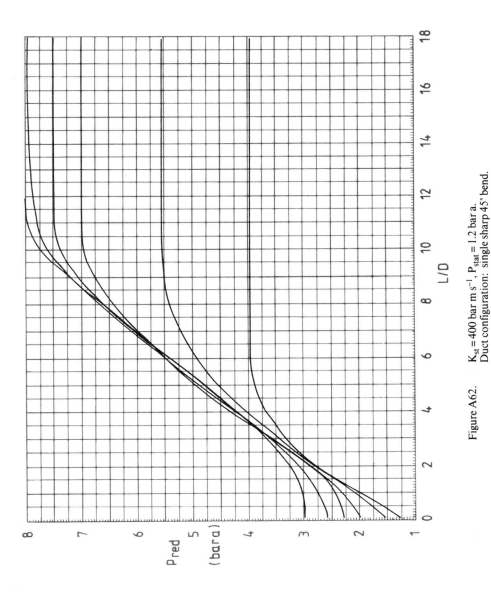

Figure A62. $K_{st} = 400$ bar m s^{-1}, $P_{stat} = 1.2$ bar a. Duct configuration: single sharp 45° bend.

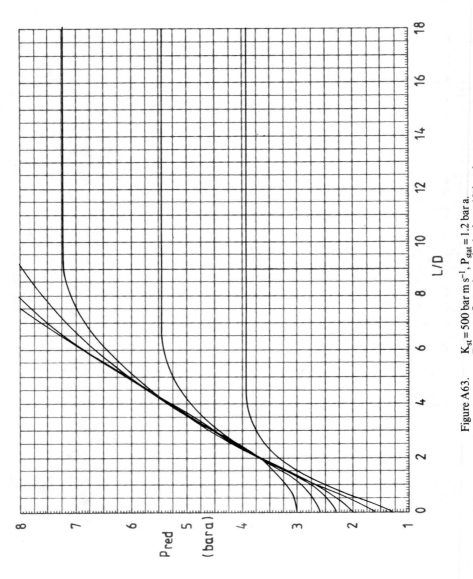

Figure A63. $K_{st} = 500$ bar m s^{-1}, $P_{stat} = 1.2$ bar a. Duct configuration: single sharp 45° bend.

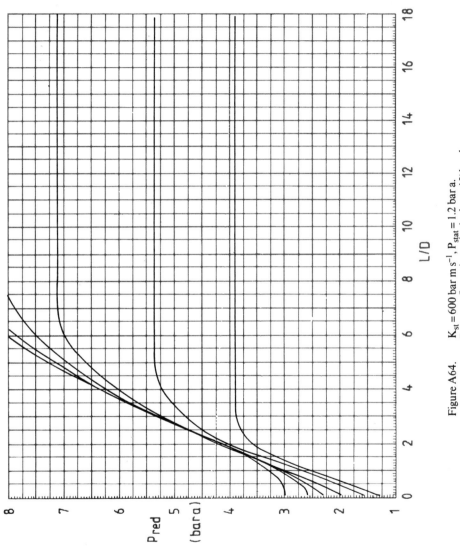

Figure A64. $K_{st} = 600$ bar m s^{-1}, $P_{stat} = 1.2$ bar a.
Duct configuration: single sharp 45° bend.

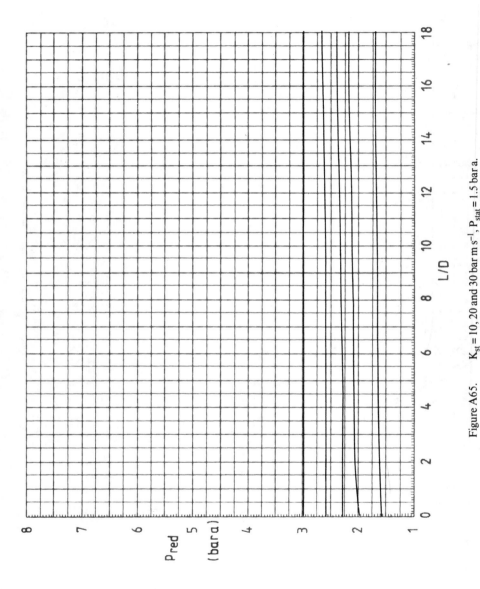

Figure A65. $K_{st} = 10, 20$ and $30\,\text{bar m s}^{-1}$, $P_{stat} = 1.5\,\text{bar a}$. Duct configuration: single sharp 45° bend.

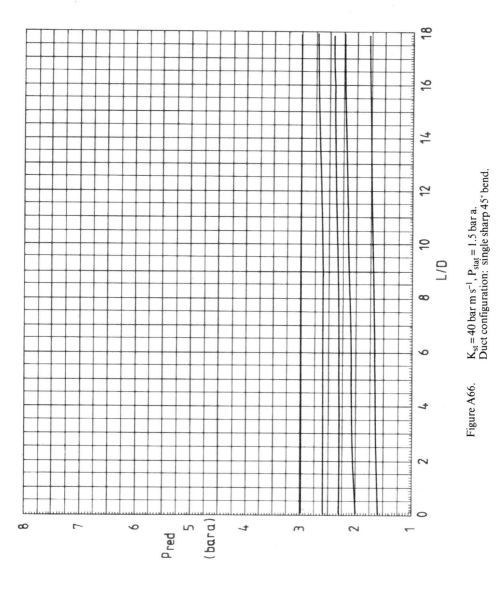

Figure A66. $K_{st} = 40$ bar m s^{-1}, $P_{stat} = 1.5$ bar a. Duct configuration: single sharp 45° bend.

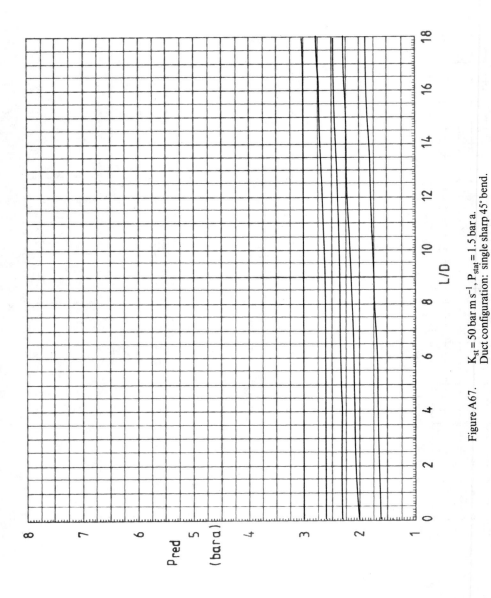

Figure A67. $K_{st} = 50$ bar m s^{-1}, $P_{stat} = 1.5$ bar a. Duct configuration: single sharp 45° bend.

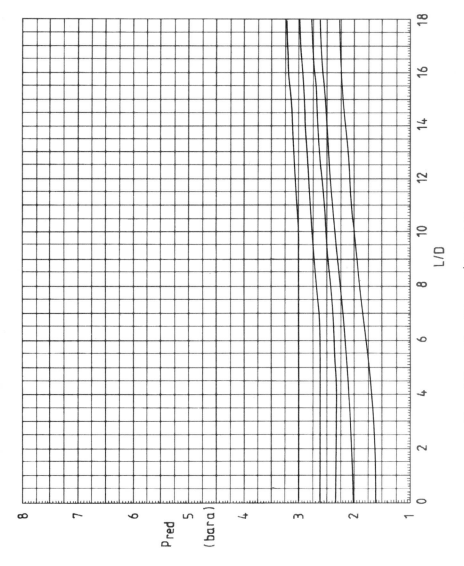

Figure A68. $K_{st} = 75$ bar m s^{-1}, $P_{stat} = 1.5$ bar a. Duct configuration: single sharp 45° bend.

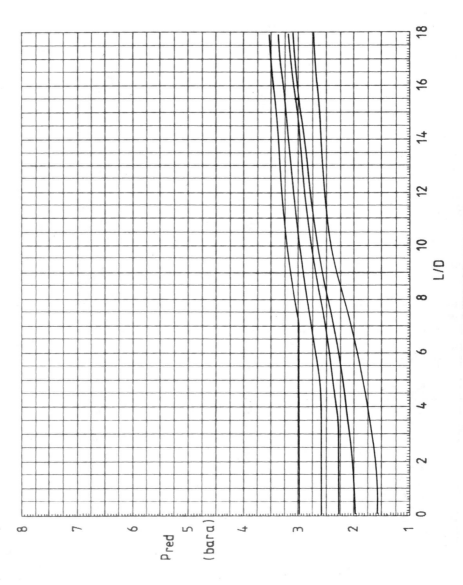

Figure A69. $K_{st} = 100$ bar m s^{-1}, $P_{stat} = 1.5$ bar a. Duct configuration: single sharp 45° bend.

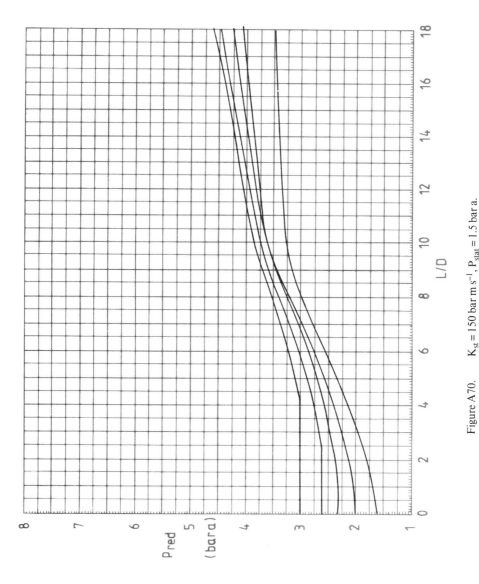

Figure A70. $K_{st} = 150$ bar m s^{-1}, $P_{stat} = 1.5$ bar a.
Duct configuration: single sharp 45° bend.

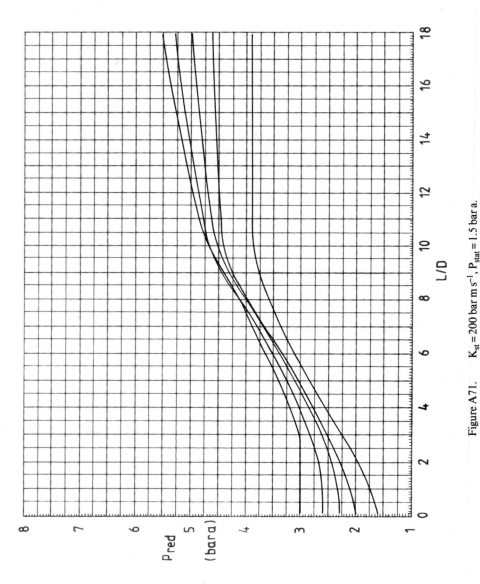

Figure A71. $K_{st} = 200 \text{ bar m s}^{-1}$, $P_{stat} = 1.5 \text{ bar a}$. Duct configuration: single sharp 45° bend.

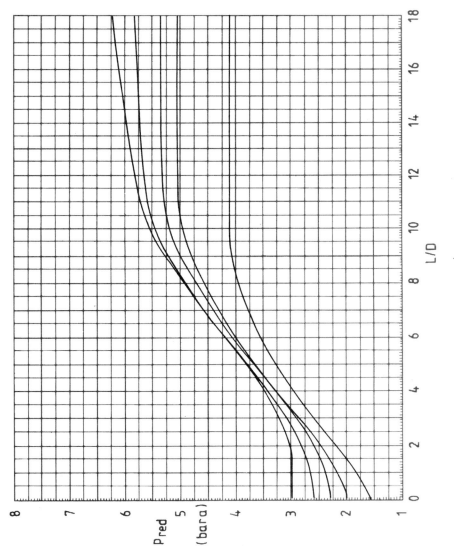

Figure A72. $K_{st} = 250$ bar m s^{-1}, $P_{stat} = 1.5$ bar a. Duct configuration: single sharp 45° bend.

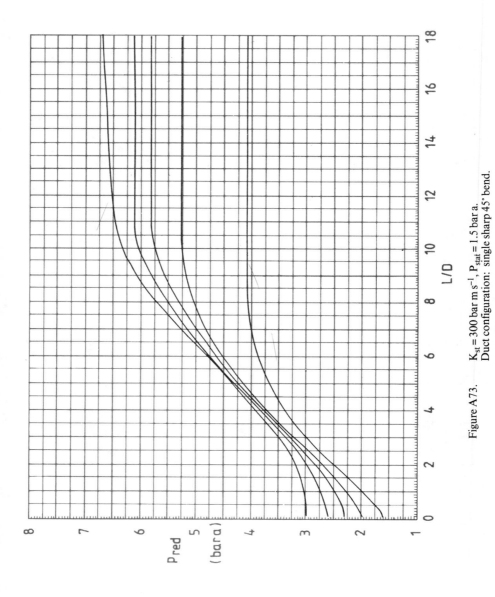

Figure A73. —— $K_{st} = 300$ bar m s^{-1}, $P_{stat} = 1.5$ bar a. Duct configuration: single sharp 45° bend.

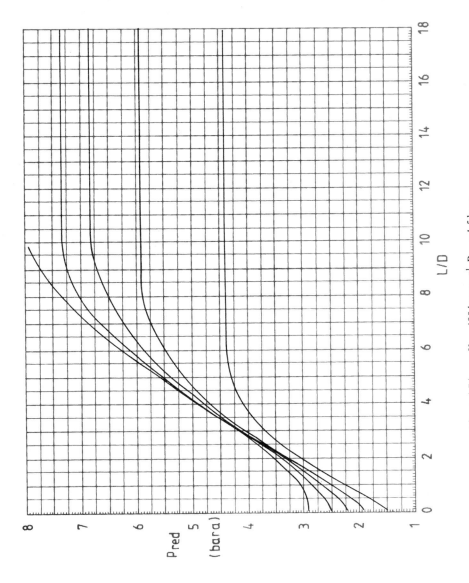

Figure A74. $K_{st} = 400$ bar m s^{-1}, $P_{stat} = 1.5$ bar a.
Duct configuration: single sharp 45° bend.

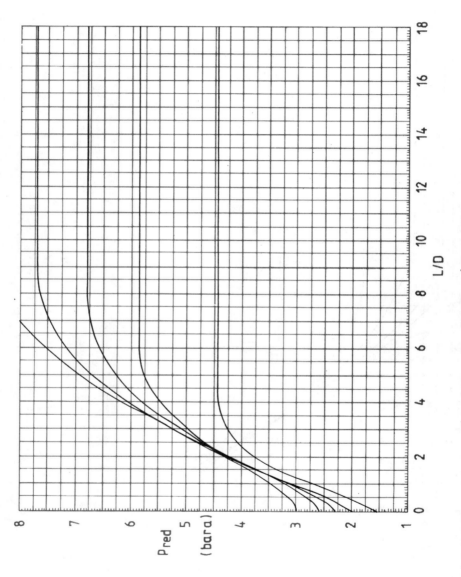

Figure A75. $K_{st} = 500$ bar m s^{-1}, $P_{stat} = 1.5$ bar a. Duct configuration: single sharp 45° bend.

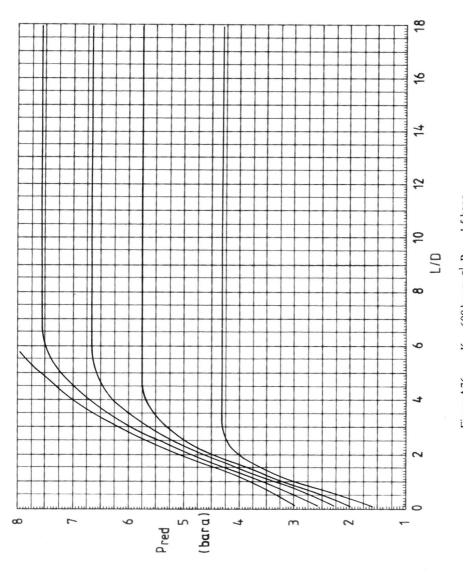

Figure A76. $K_{st} = 600$ bar m s^{-1}, $P_{stat} = 1.5$ bar a. Duct configuration: single sharp 45° bend.

APPENDIX 3

ESTIMATES OF REDUCED EXPLOSION PRESSURES FOR VENT DUCTS WITH A SINGLE SHARP 90° BEND

NOTE: All pressures are given in bar a.

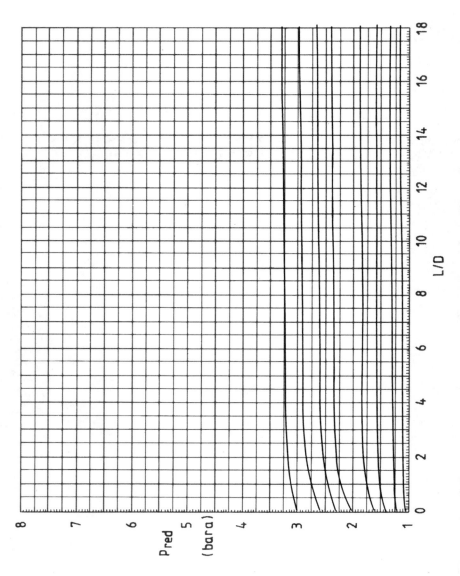

Figure A77. $K_{st} = 10$ and 20 bar m s^{-1}, $P_{stat} = 1.1$ bar a. Duct configuration: single sharp 90° bend.

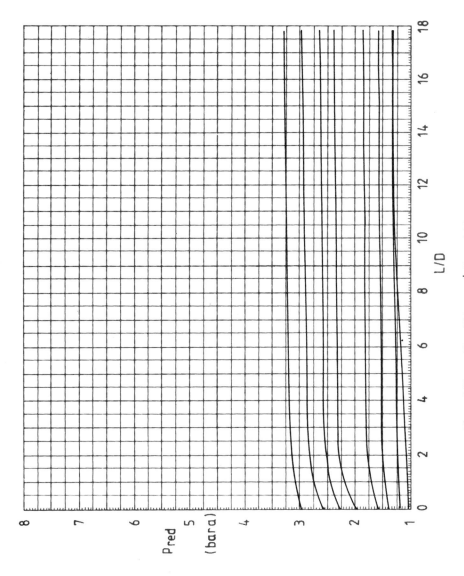

Figure A78. $K_{st} = 30$ bar m s^{-1}, $P_{stat} = 1.1$ bar a. Duct configuration: single sharp 90° bend.

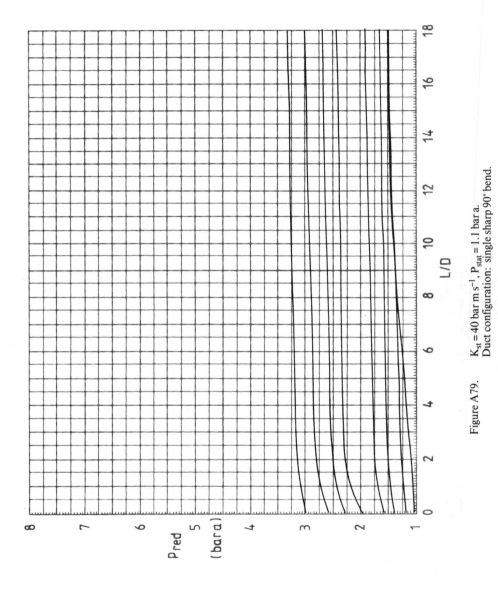

Figure A79. $K_{st} = 40$ bar m s^{-1}, $P_{stat} = 1.1$ bar a.
Duct configuration: single sharp 90° bend.

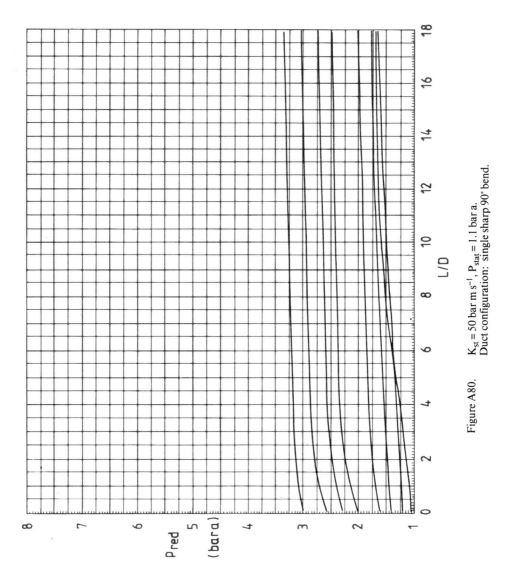

Figure A80. K_{st} = 50 bar m s^{-1}, P_{stat} = 1.1 bar a. Duct configuration: single sharp 90° bend.

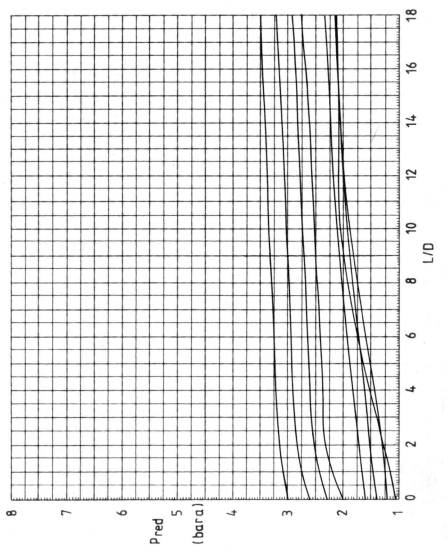

Figure A81. $K_{st} = 75$ bar m s^{-1}, $P_{stat} = 1.1$ bar a. Duct configuration: single sharp 90° bend.

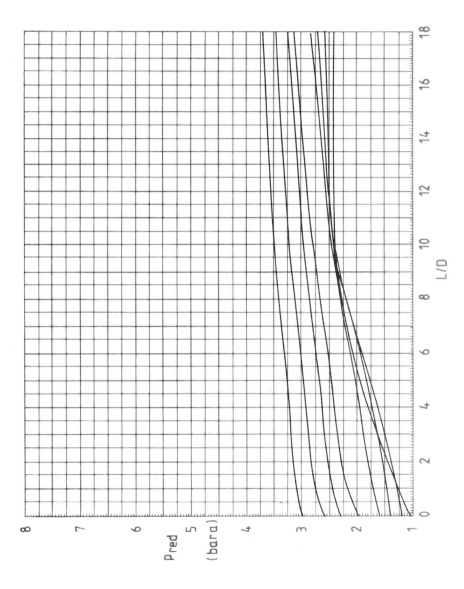

Figure A82. $K_{st} = 100$ bar m s^{-1}, $P_{stat} = 1.1$ bar a. Duct configuration: single sharp 90° bend.

A GUIDE TO DUST EXPLOSION, PART 3

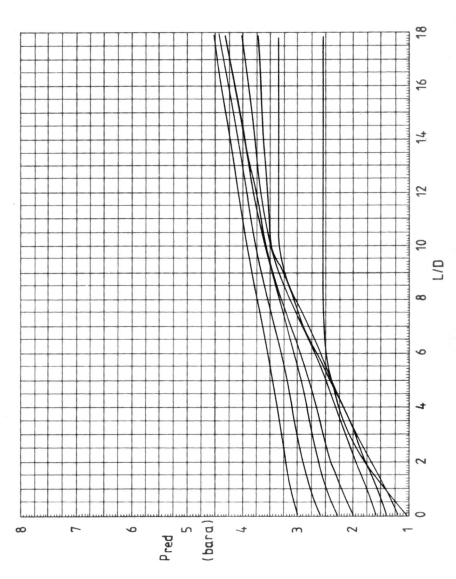

Figure A83. $K_{st} = 150$ bar m s^{-1}, $P_{stat} = 1.1$ bar a.
Duct configuration: single sharp 90° bend.

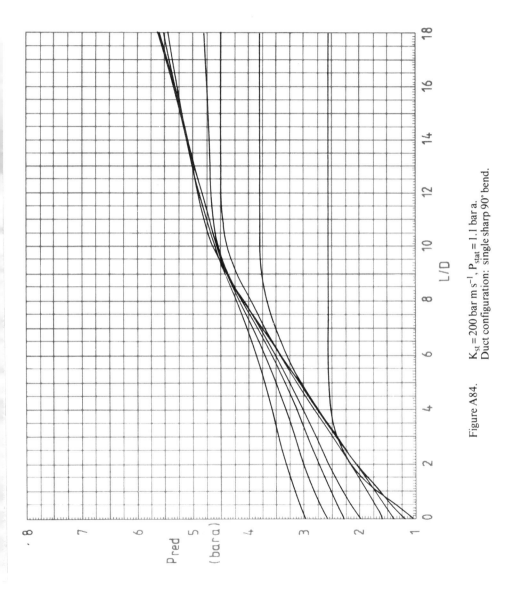

Figure A84. $K_{st} = 200$ bar m s^{-1}, $P_{stat} = 1.1$ bar a. Duct configuration: single sharp 90° bend.

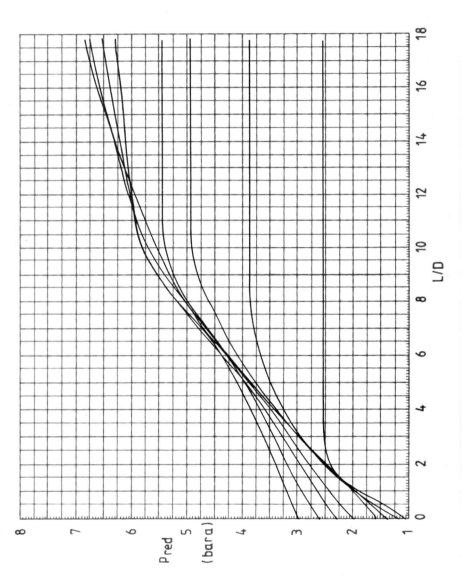

Figure A85. $K_{st} = 250$ bar m s^{-1}, $P_{stat} = 1.1$ bar a.
Duct configuration: single sharp 90° bend.

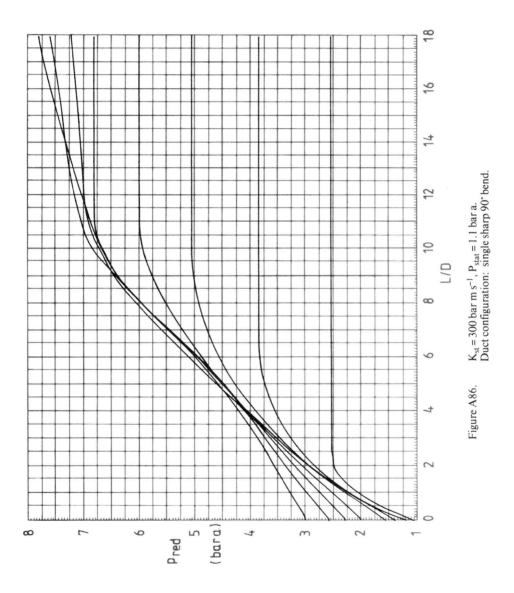

Figure A86. $K_{st} = 300$ bar m s^{-1}, $P_{stat} = 1.1$ bar a.
Duct configuration: single sharp 90° bend.

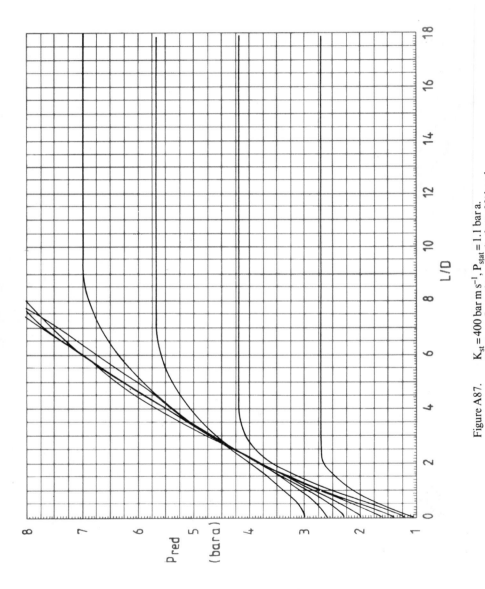

Figure A87. $K_{st} = 400$ bar m s^{-1}, $P_{stat} = 1.1$ bar a. Duct configuration: single sharp 90° bend.

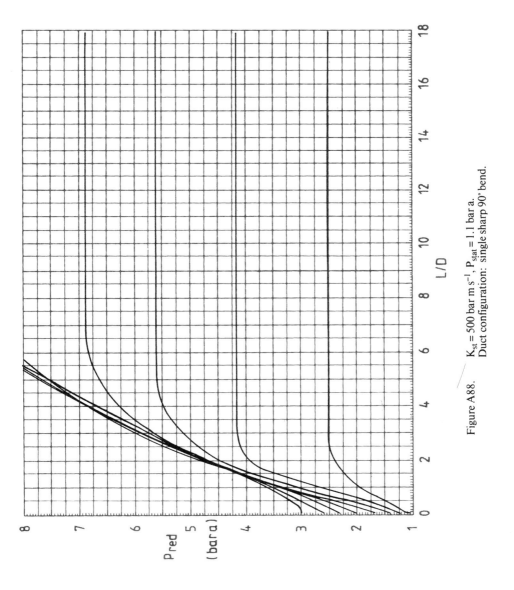

Figure A88. $K_{st} = 500$ bar m s^{-1}, $P_{stat} = 1.1$ bar a. Duct configuration: single sharp 90° bend.

A GUIDE TO DUST EXPLOSION, PART 3

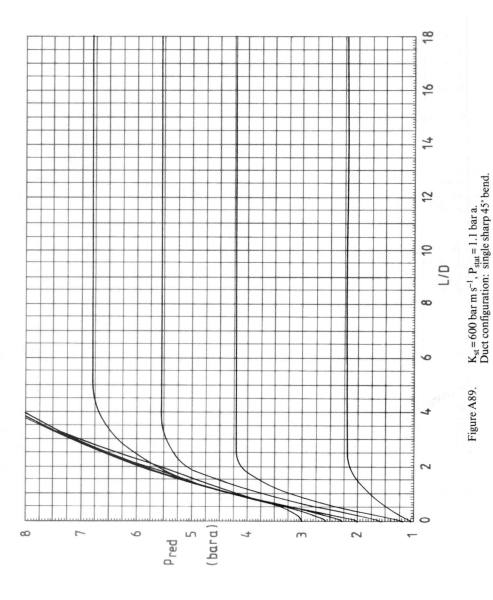

Figure A89. $K_{st} = 600$ bar m s^{-1}, $P_{stat} = 1.1$ bar a. Duct configuration: single sharp 45° bend.

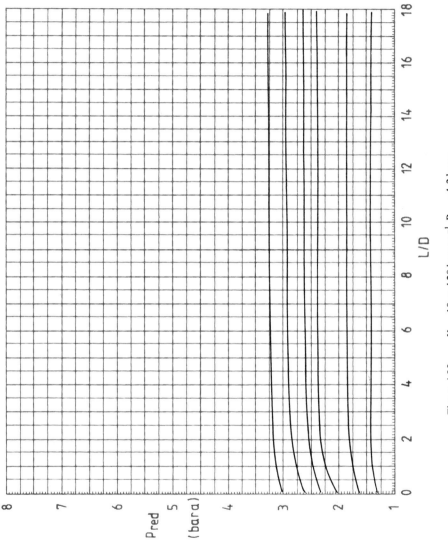

Figure A90. $K_{st} = 10$ and 20 bar m s^{-1}, $P_{stat} = 1.2$ bar a. Duct configuration: single sharp 90° bend.

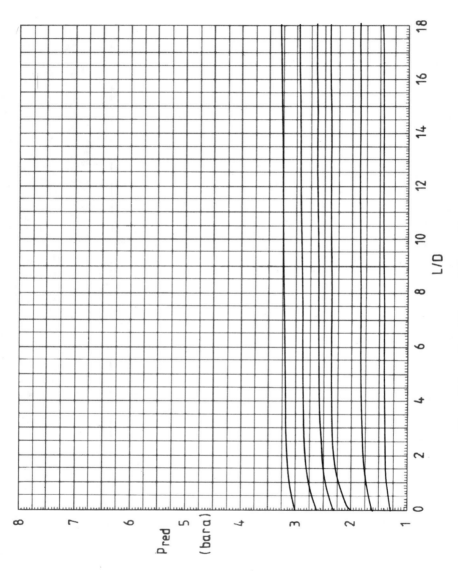

Figure A91. $K_{st} = 30$ bar m s^{-1}, $P_{stat} = 1.2$ bar a.
Duct configuration: single sharp 90° bend.

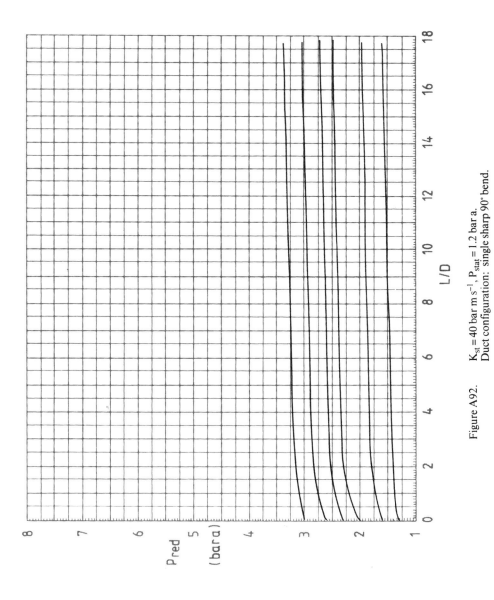

Figure A92. $K_{st} = 40$ bar m s^{-1}, $P_{stat} = 1.2$ bar a.
Duct configuration: single sharp 90° bend.

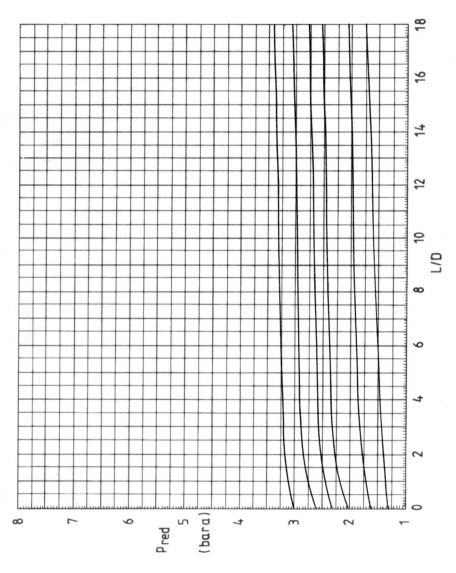

Figure A93. $K_{st} = 50$ bar m s^{-1}, $P_{stat} = 1.2$ bar a. Duct configuration: single sharp 90° bend.

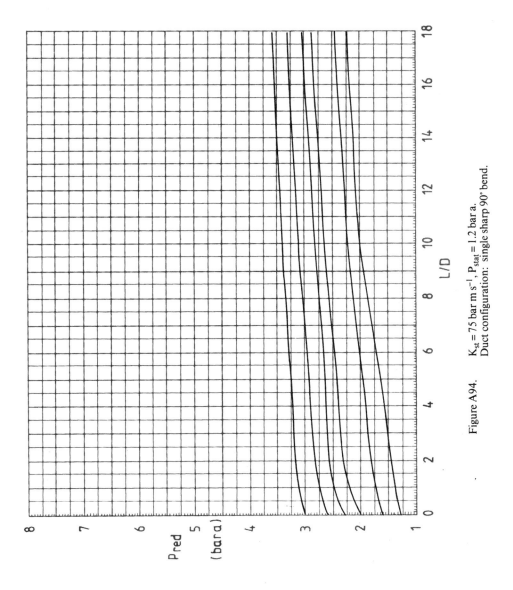

Figure A94. $K_{st} = 75$ bar m s^{-1}, $P_{stat} = 1.2$ bar a. Duct configuration: single sharp 90° bend.

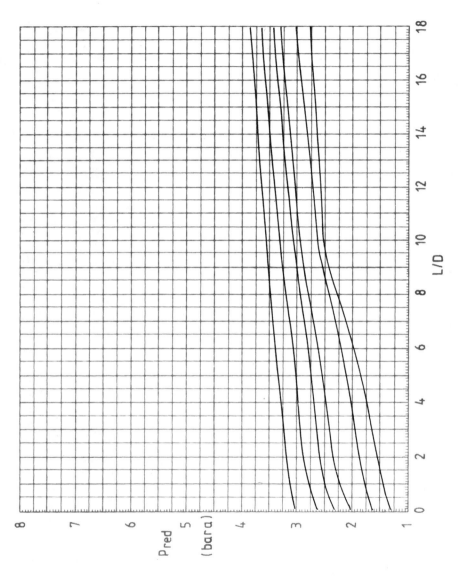

Figure A95. $K_{st} = 100$ bar m s^{-1}, $P_{stat} = 1.2$ bar a. Duct configuration: single sharp 90° bend.

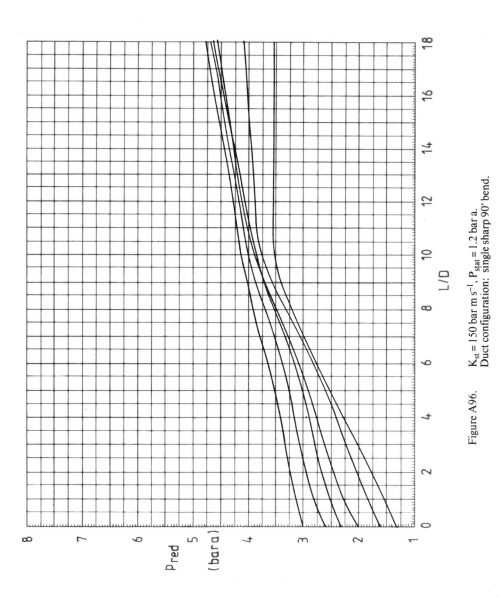

Figure A96. $K_{st} = 150$ bar m s^{-1}, $P_{stat} = 1.2$ bar a. Duct configuration: single sharp 90° bend.

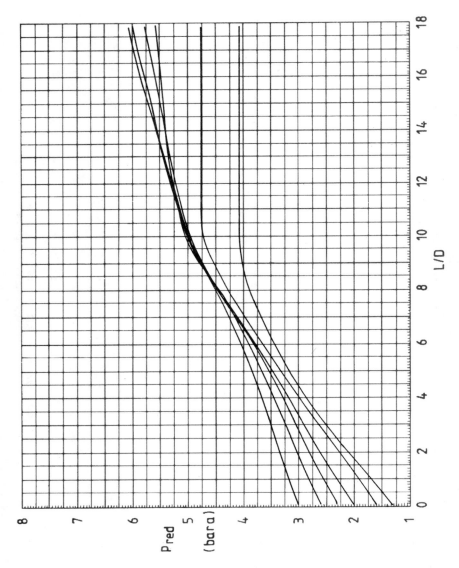

Figure A97. $K_{st} = 200$ bar m s^{-1}, $P_{stat} = 1.2$ bar a. Duct configuration: single sharp 90° bend.

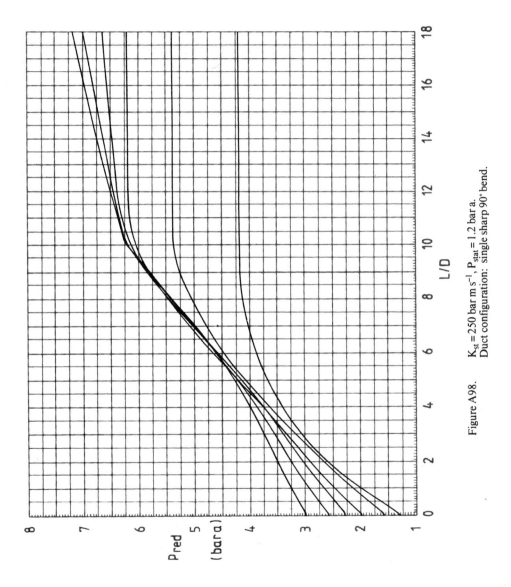

Figure A98. $K_{st} = 250$ bar m s^{-1}, $P_{stat} = 1.2$ bar a. Duct configuration: single sharp 90° bend.

A GUIDE TO DUST EXPLOSION, PART 3

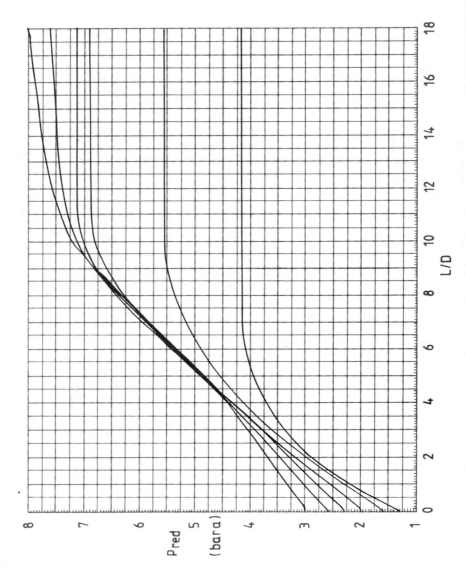

Figure A99. $K_{st} = 300$ bar m s^{-1}, $P_{stat} = 1.2$ bar a.
Duct configuration: single sharp 90° bend.

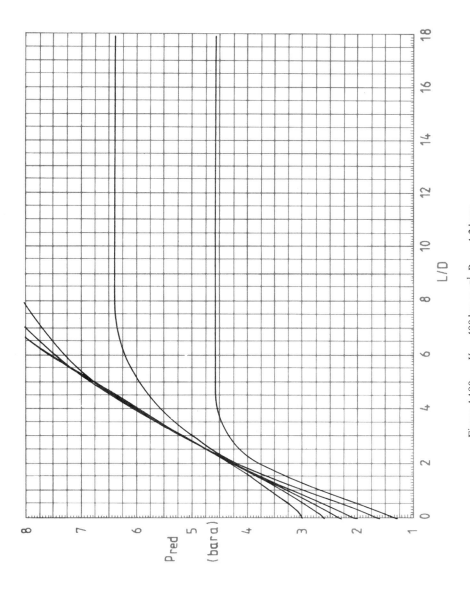

Figure A100. $K_{st} = 400$ bar m s^{-1}, $P_{stat} = 1.2$ bar a. Duct configuration: single sharp 90° bend.

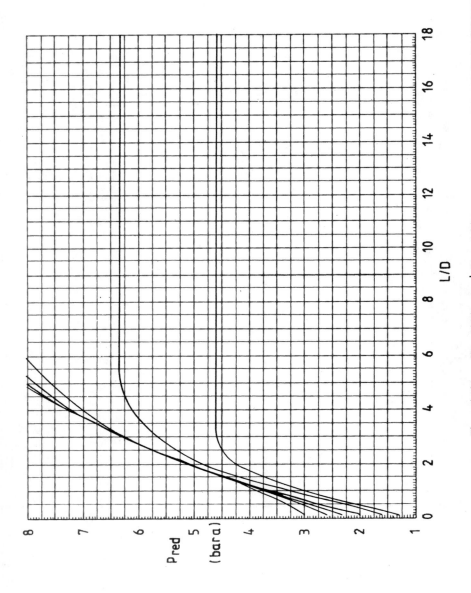

Figure A101. $K_{st} = 500$ bar m s^{-1}, $P_{stat} = 1.2$ bar a. Duct configuration: single sharp 90° bend.

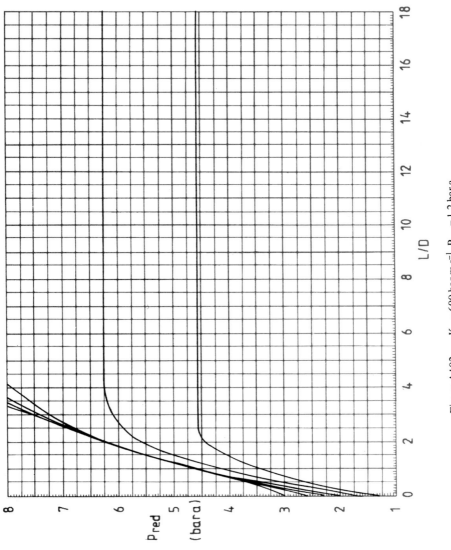

Figure A102. $K_{st} = 600$ bar m s^{-1}, $P_{stat} = 1.2$ bar a. Duct configuration: single sharp 90° bend.

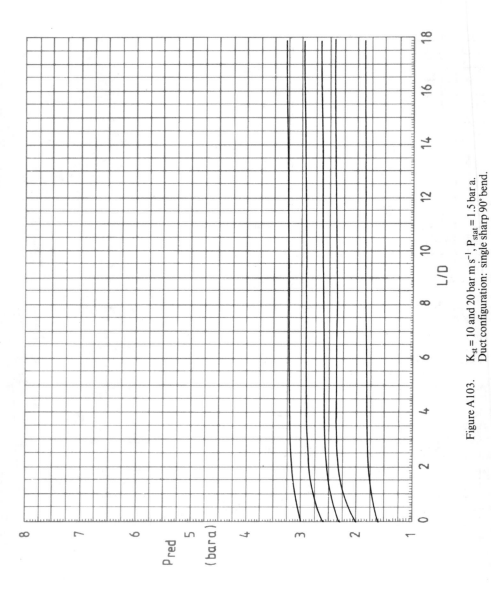

Figure A103. $K_{st} = 10$ and $20\,\text{bar m s}^{-1}$, $P_{stat} = 1.5\,\text{bar a}$. Duct configuration: single sharp 90° bend.

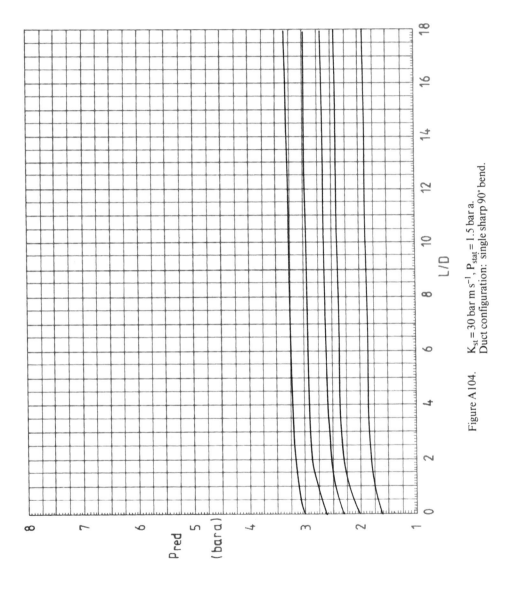

Figure A104. $K_{st} = 30$ bar m s^{-1}, $P_{stat} = 1.5$ bar a.
Duct configuration: single sharp 90° bend.

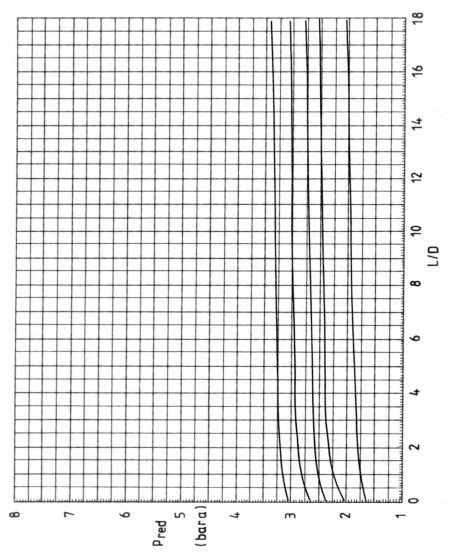

Figure A105. $K_{st} = 40$ bar m s^{-1}, $P_{stat} = 1.5$ bar a. Duct configuration: single sharp 90° bend.

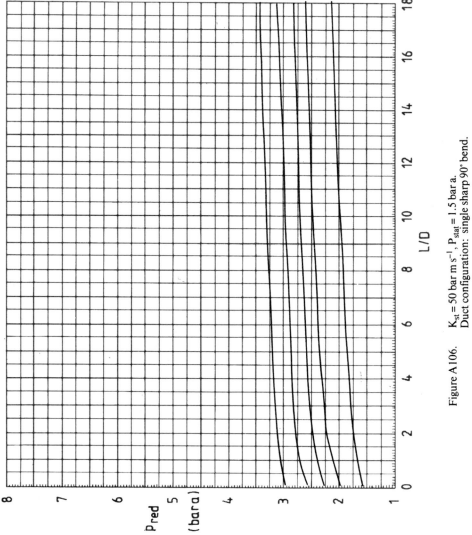

Figure A106. $K_{st} = 50$ bar m s^{-1}, $P_{stat} = 1.5$ bar a. Duct configuration: single sharp 90° bend.

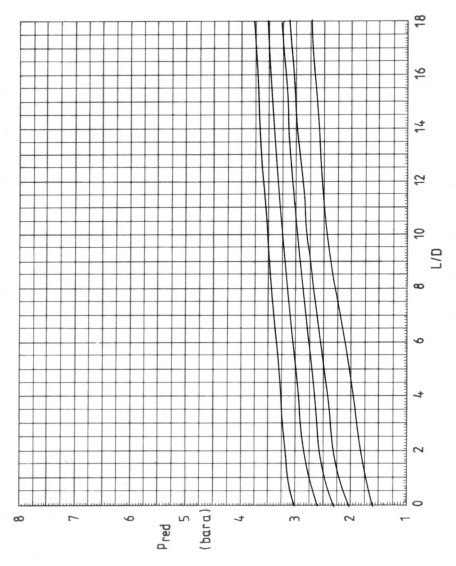

Figure A107. $K_{st} = 75$ bar m s^{-1}, $P_{stat} = 1.5$ bar a.
Duct configuration: single sharp 90° bend.

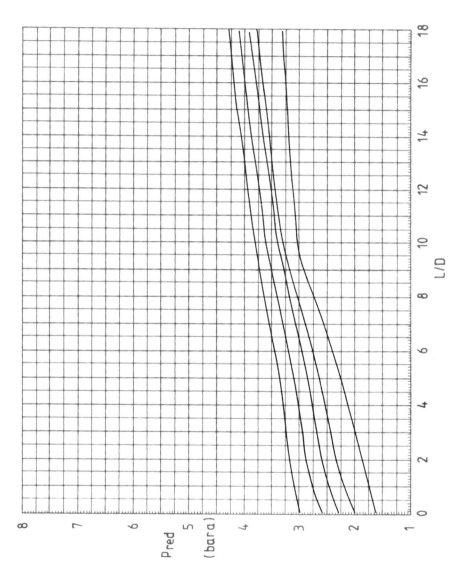

Figure A108. $K_{st} = 100$ bar m s^{-1}, $P_{stat} = 1.5$ bar a. Duct configuration: single sharp 90° bend.

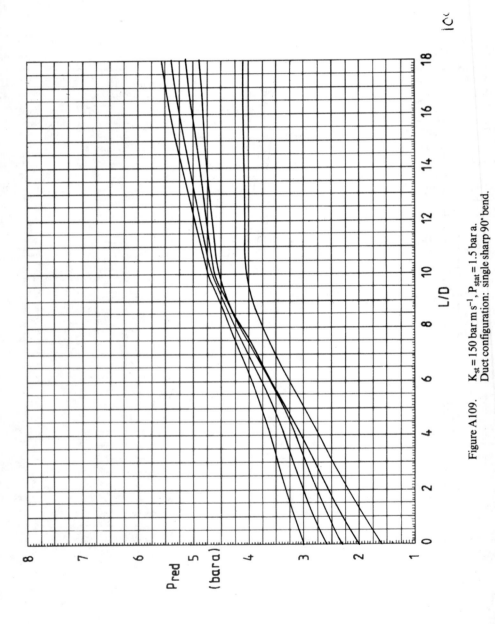

Figure A109. $K_{st} = 150$ bar m s^{-1}, $P_{stat} = 1.5$ bar a. Duct configuration: single sharp 90° bend.

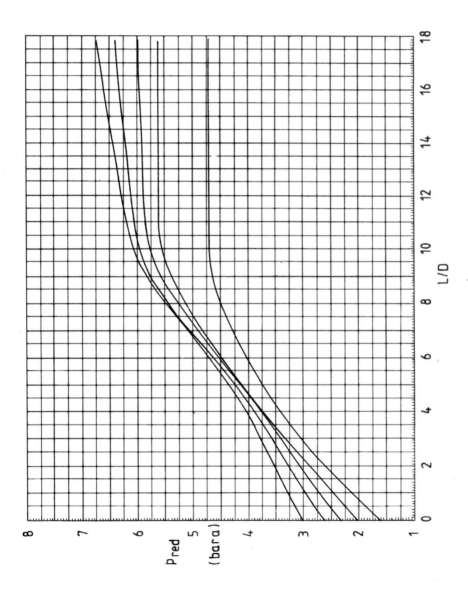

Figure A110. $K_{st} = 200$ bar m s^{-1}, $P_{stat} = 1.5$ bar a.
Duct configuration: single sharp 90° bend.

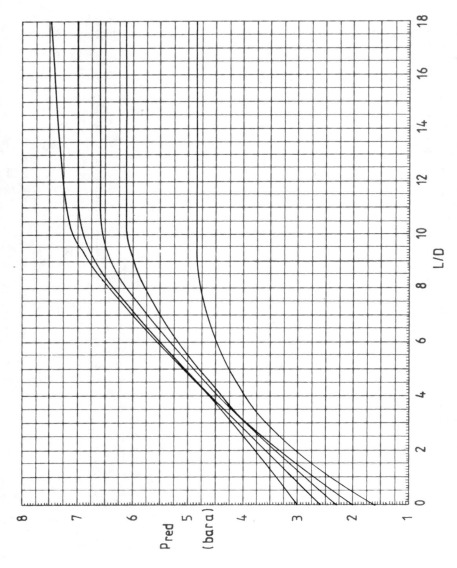

Figure A111. $K_{st} = 250$ bar m s^{-1}, $P_{stat} = 1.5$ bar a. Duct configuration: single sharp 90° bend.

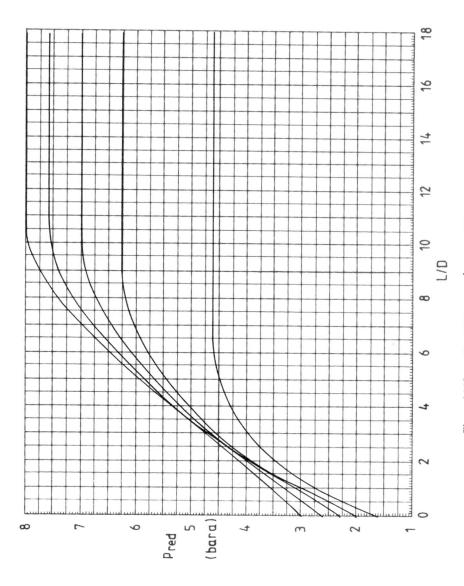

Figure A112. $K_{st} = 300$ bar m s^{-1}, $P_{stat} = 1.5$ bar a. Duct configuration: single sharp 90° bend.

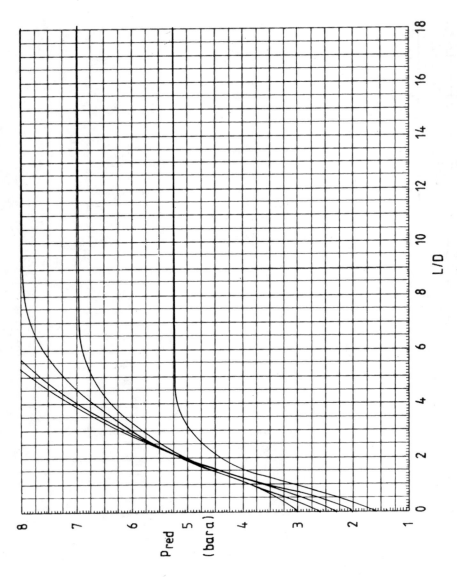

Figure A113. $K_{st} = 400$ bar m s^{-1}, $P_{stat} = 1.5$ bar a. Duct configuration: single sharp 90° bend.

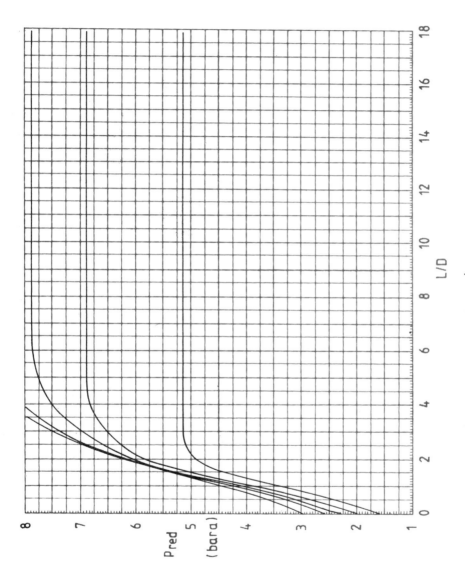

Figure A114. $K_{st} = 500$ bar m s^{-1}, $P_{stat} = 1.5$ bar a. Duct configuration: single sharp 90° bend.

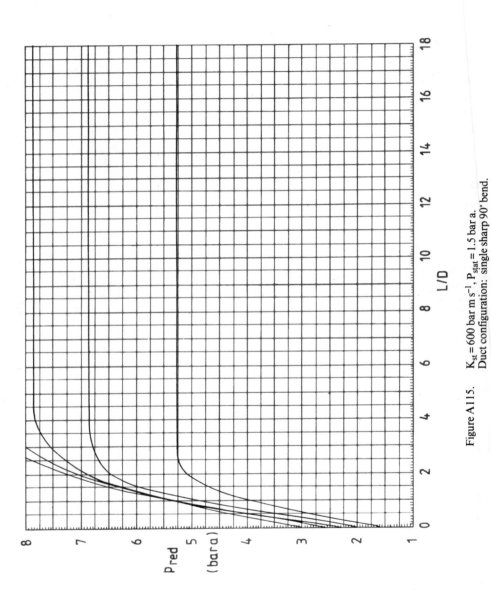

Figure A115. $K_{st} = 600$ bar m s^{-1}, $P_{stat} = 1.5$ bar a. Duct configuration: single sharp 90° bend.

APPENDIX 4

ESTIMATES OF THE REDUCED EXPLOSION PRESSURES FOR St3 METAL DUSTS

NOTE: All pressures are given in bar a.

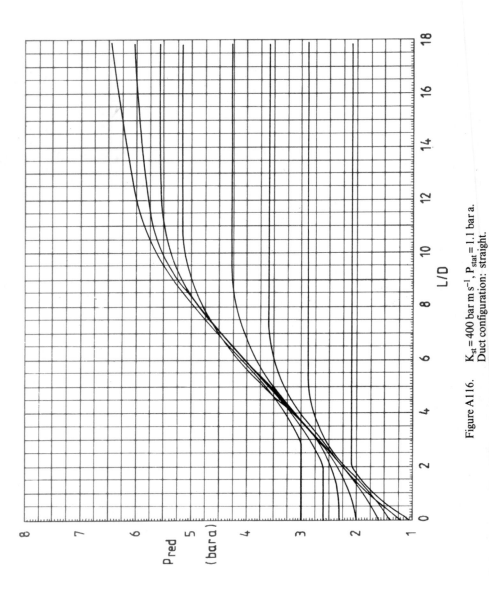

Figure A116. $K_{st} = 400$ bar m s^{-1}, $P_{stat} = 1.1$ bar a. Duct configuration: straight.

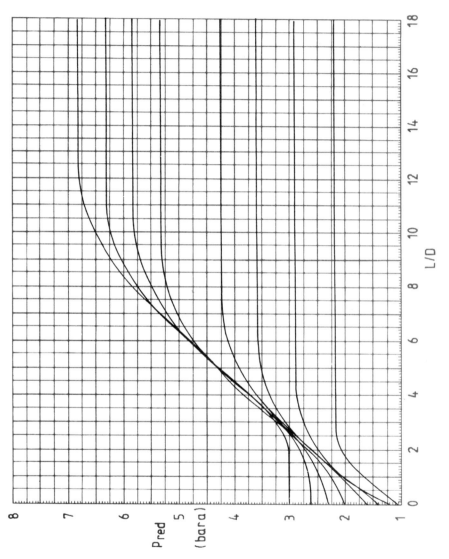

Figure A117. $K_{st} = 500$ bar m s^{-1}, $P_{stat} = 1.1$ bar a. Duct configuration: straight.

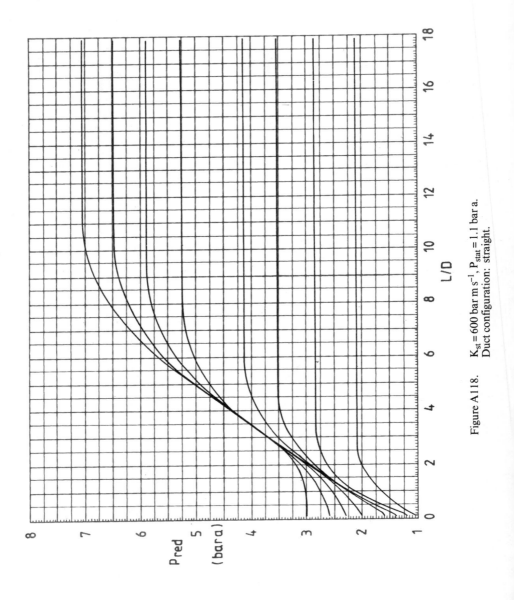

Figure A118. $K_{st} = 600$ bar m s^{-1}, $P_{stat} = 1.1$ bar a. Duct configuration: straight.

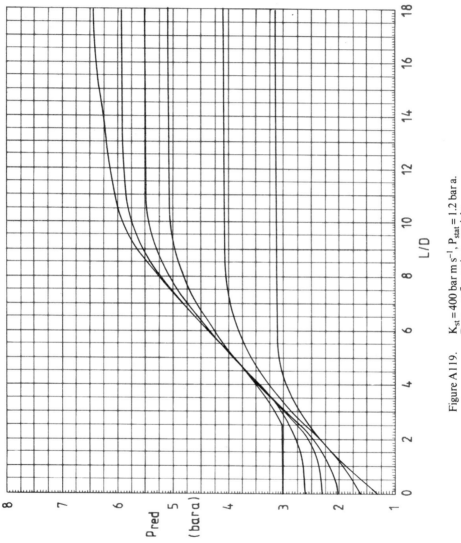

Figure A119. $K_{st} = 400$ bar m s^{-1}, $P_{stat} = 1.2$ bar a. Duct configuration: straight.

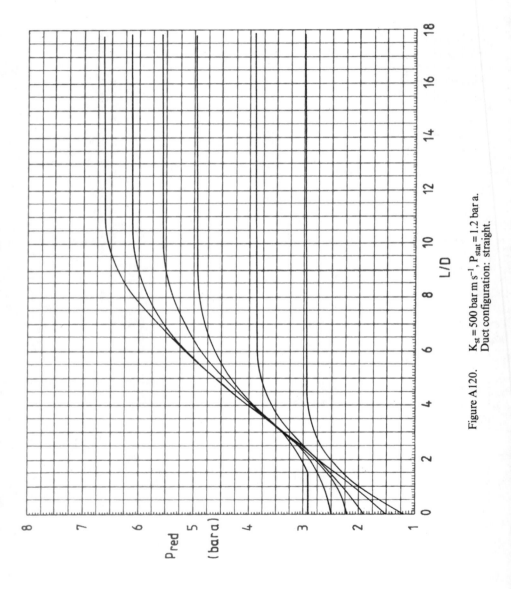

Figure A120. K_{st} = 500 bar m s^{-1}, P_{stat} = 1.2 bar a. Duct configuration: straight.

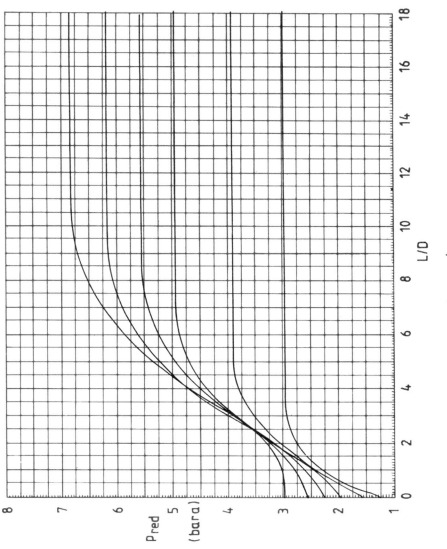

Figure A121. $K_{st} = 600$ bar m s^{-1}, $P_{stat} = 1.2$ bar a. Duct configuration: straight.

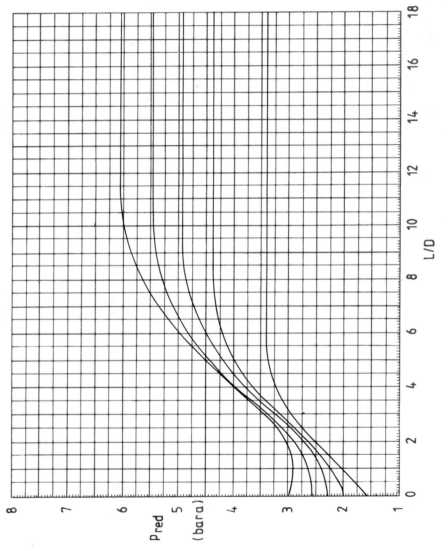

Figure A122. $K_{st} = 400$ bar m s^{-1}, $P_{stat} = 1.5$ bar a. Duct configuration: straight.

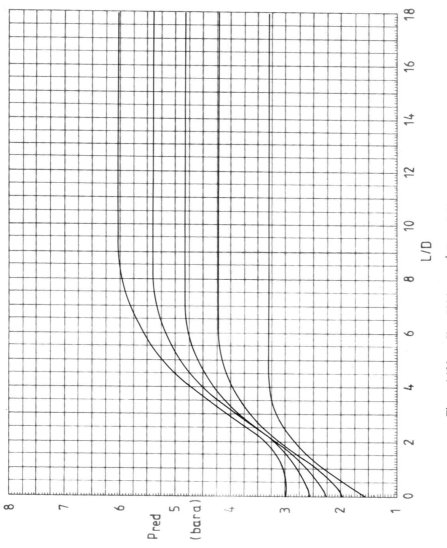

Figure A123. $K_{st} = 500$ bar m s^{-1}, $P_{stat} = 1.5$ bar a. Duct configuration: straight.

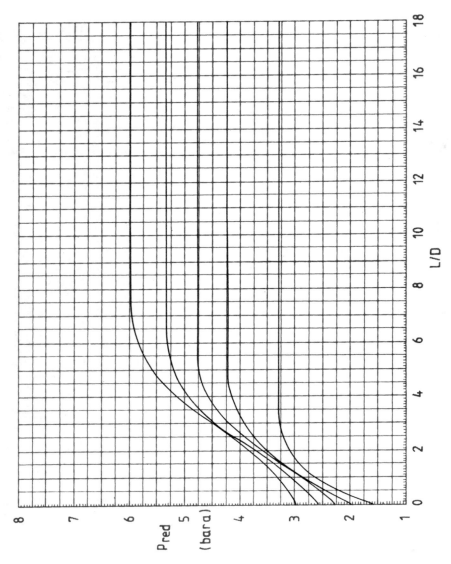

Figure A124. $K_{st} = 600$ bar m s^{-1}, $P_{stat} = 1.5$ bar a. Duct configuration: straight.

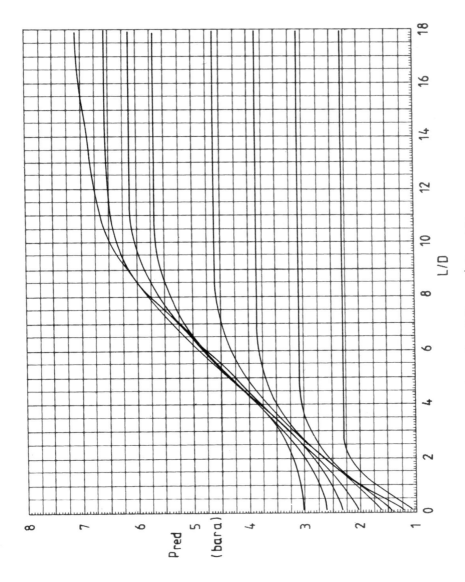

Figure A125. $K_{st} = 400$ bar m s^{-1}, $P_{stat} = 1.1$ bar a.
Duct configuration: single sharp 45° bend.

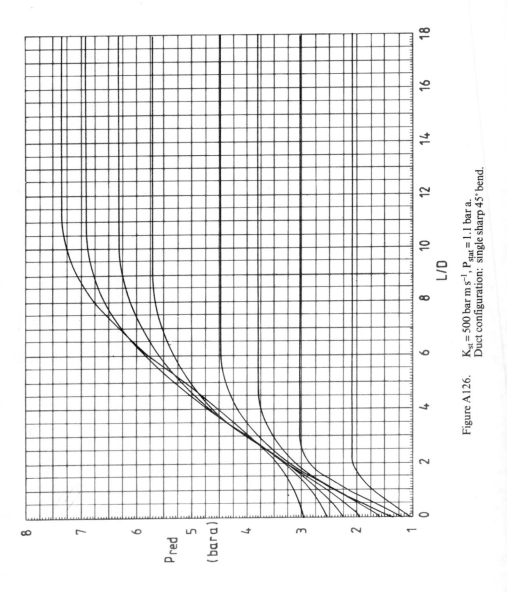

Figure A126. $K_{st} = 500$ bar m s^{-1}, $P_{stat} = 1.1$ bar a. Duct configuration: single sharp 45° bend.

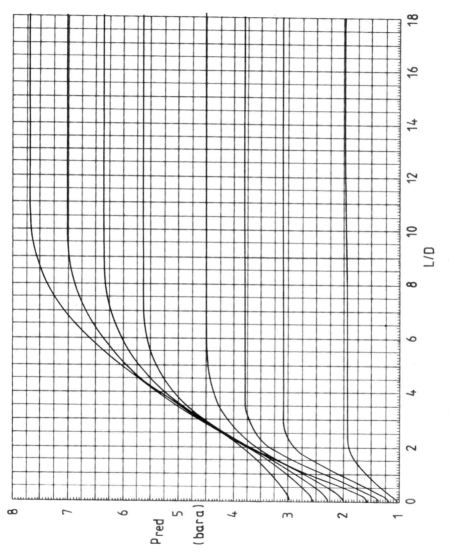

Figure A127. $K_{st} = 600$ bar m s^{-1}, $P_{stat} = 1.1$ bar a. Duct configuration: single sharp 45° bend.

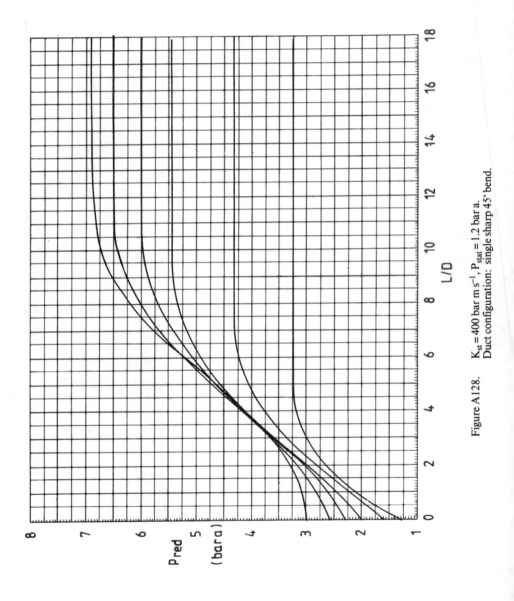

Figure A128. $K_{st} = 400$ bar m s^{-1}, $P_{stat} = 1.2$ bar a. Duct configuration: single sharp 45° bend.

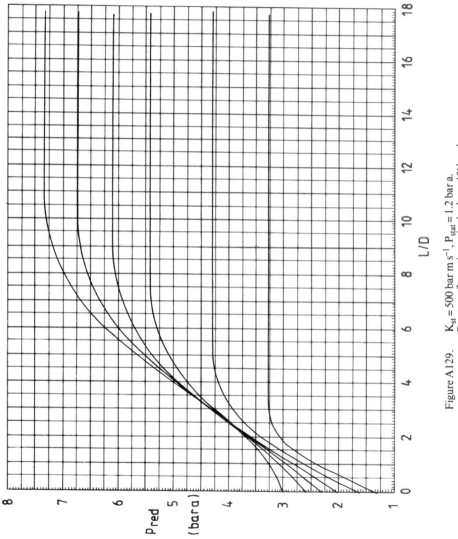

Figure A129. $K_{st} = 500$ bar m s^{-1}, $P_{stat} = 1.2$ bar a. Duct configuration: single sharp 45° bend.

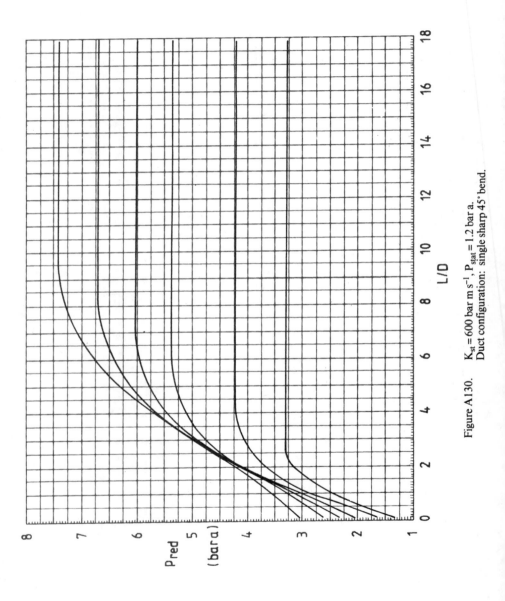

Figure A130. $K_{st} = 600$ bar m s^{-1}, $P_{stat} = 1.2$ bar a. Duct configuration: single sharp 45° bend.

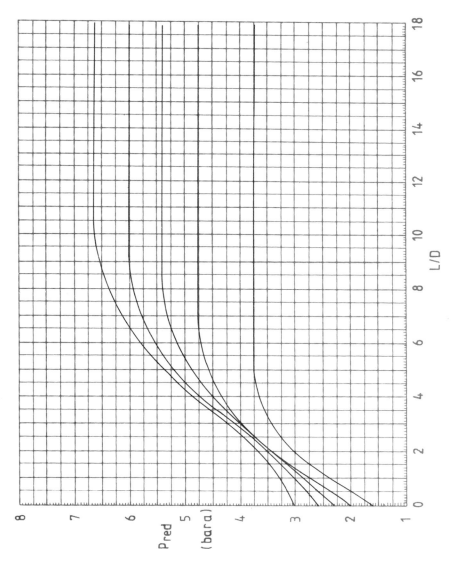

Figure A131. $K_{st} = 400$ bar m s^{-1}, $P_{stat} = 1.5$ bar a. Duct configuration: single sharp 45° bend.

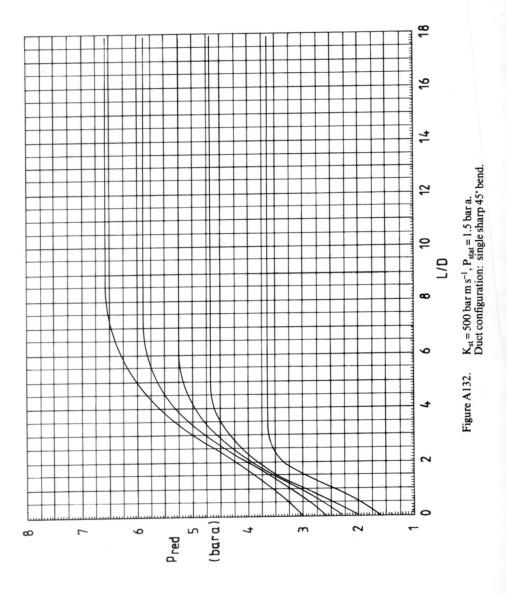

Figure A132. $K_{st} = 500\,\text{bar m s}^{-1}$, $P_{stat} = 1.5\,\text{bar a}$. Duct configuration: single sharp 45° bend.

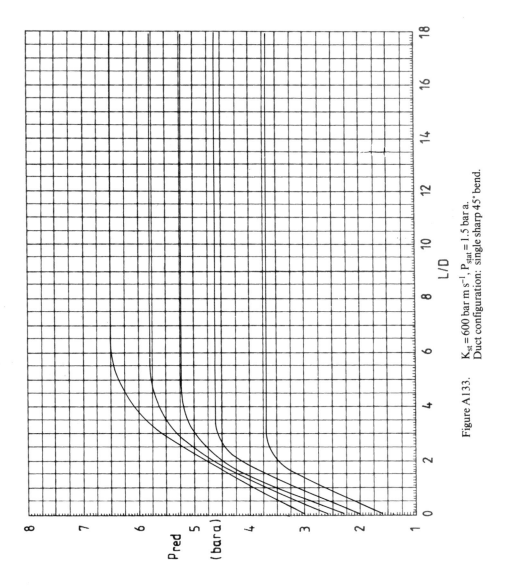

Figure A133. $K_{st} = 600$ bar m s^{-1}, $P_{stat} = 1.5$ bar a. Duct configuration: single sharp 45° bend.

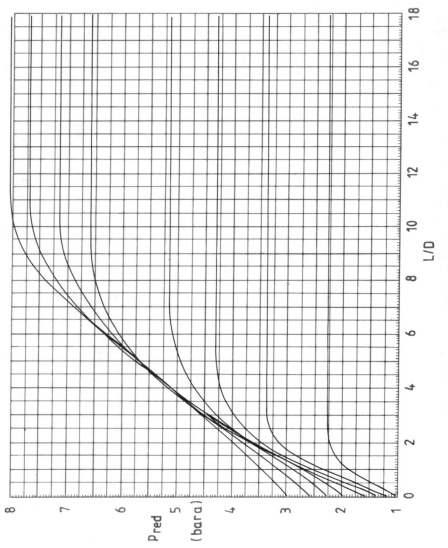

Figure A134. $K_{st} = 400$ bar m s^{-1}, $P_{stat} = 1.1$ bar a.
Duct configuration: single sharp 90° bend.

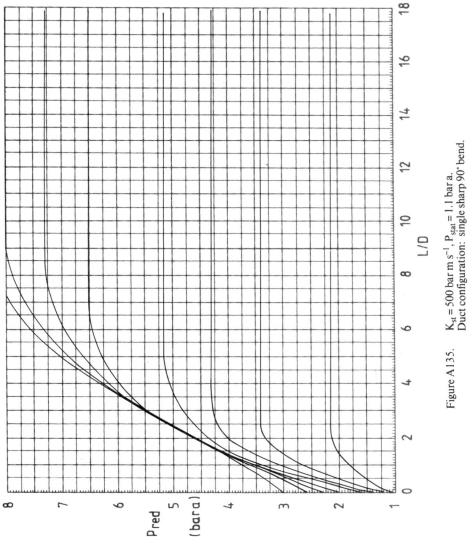

Figure A135. $K_{st} = 500$ bar m s^{-1}, $P_{stat} = 1.1$ bar a. Duct configuration: single sharp 90° bend.

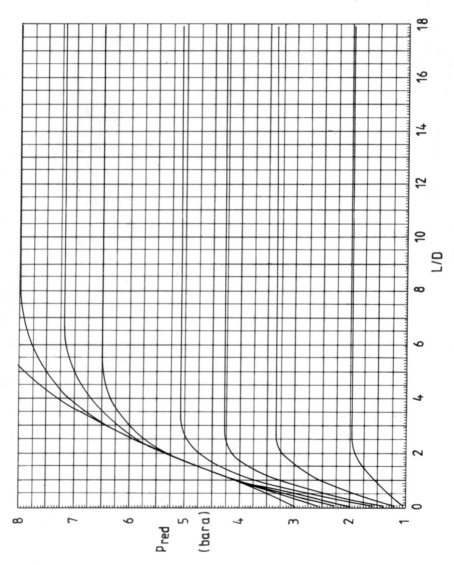

Figure A136. $K_{st} = 600$ bar m s^{-1}, $P_{stat} = 1.1$ bar a.
Duct configuration: single sharp 90° bend.

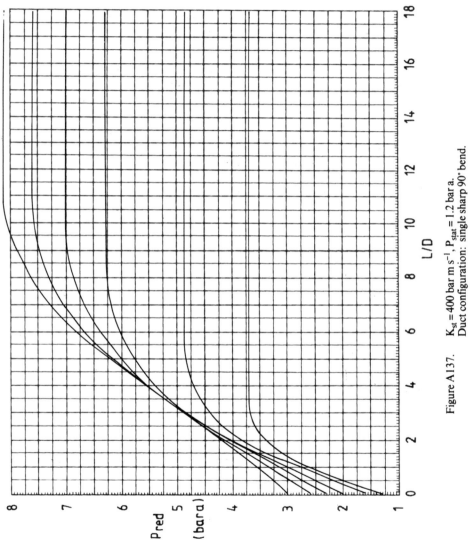

Figure A137. $K_{st} = 400$ bar m s^{-1}, $P_{stat} = 1.2$ bar a.
Duct configuration: single sharp 90° bend.

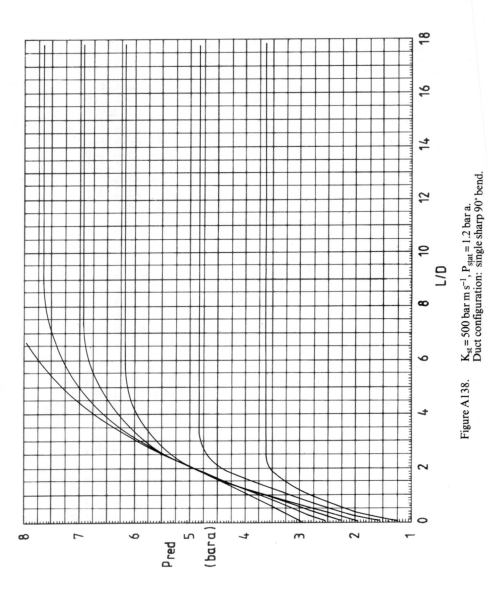

Figure A138. $K_{st} = 500$ bar m s^{-1}, $P_{stat} = 1.2$ bar a. Duct configuration: single sharp 90° bend.

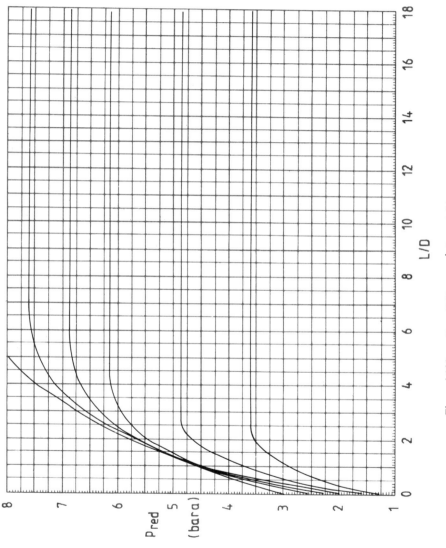

Figure A139. $K_{st} = 600$ bar m s^{-1}, $P_{stat} = 1.2$ bar a.
Duct configuration: single sharp 90° bend.

A GUIDE TO DUST EXPLOSION, PART 3

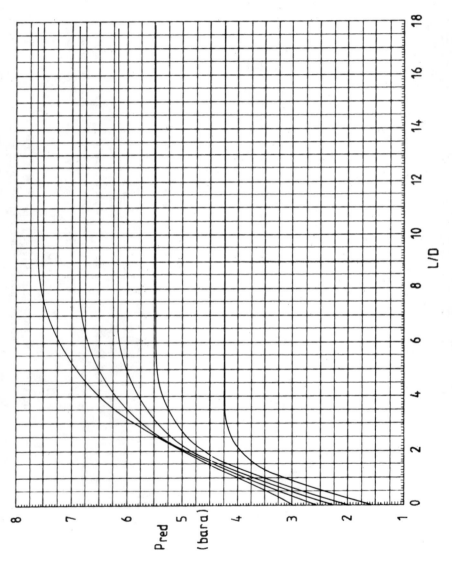

Figure A140. $K_{st} = 400$ bar m s^{-1}; $P_{stat} = 1.5$ bar a. Duct configuration: single sharp 90° bend.

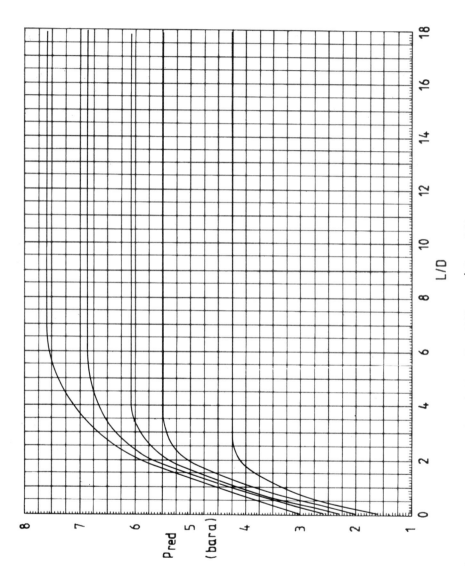

Figure A141. $K_{st} = 500$ bar m s^{-1}, $P_{stat} = 1.5$ bar a. Duct configuration: single sharp 90° bend.

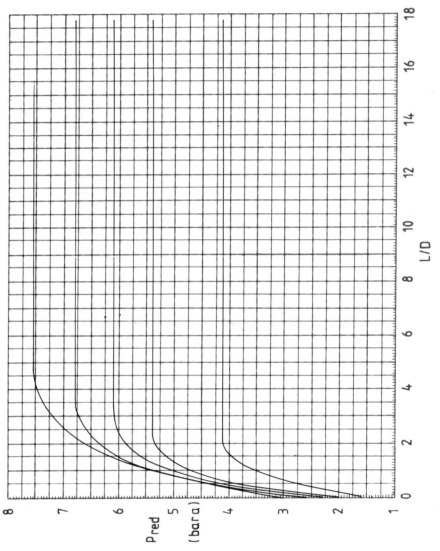

Figure A142. $K_{st} = 600$ bar m s^{-1}, $P_{stat} = 1.5$ bar a. Duct configuration: single sharp 90° bend.